Contents

Foreword

Louis H. Kauffman and Søren Brier:
Peirce and Spencer-Brown: History and synergies in cybersemiotics 3

Articles

Robin Robertson:
One, Two, Three . . . Continuity:
  C.S. Peirce and the Nature of the Continuum 7

Jack Engstrom:
Precursors to Laws of Form in C.S. Peirce's Collected Papers 25

Robin Robertson:
C.S. Peirce's "First Curiosity": The World's Most Complicated Card Trick 67

William Howard:
Peirce's Influence on Today's Mathematical Logic 69

Louis H. Kauffman:
The Mathematics of Charles Sanders Peirce 79

Inna Semetsky:
Signs in Action: Tarot as a self-organized system 111

ASC Pages

Louis H. Kauffman:
Cybernetics of Fixed Points 133

Column

Ranulph Glanville:
A (Cybernetic) Musing: Constructing My Cybernetic World 141

Reviews

Nina Ort: Searching for the Limits of Semiotics 151

Winfried Nöth: Biosemiotica 157

The artist of this issue is Jørn Særker Sørensen. Poems by Lawren Bale

Morphogenic Resonance

This concept, itself
comes back as an echo
on a noisy loop
distorted echoes
coiling back on recursive nets
where hierarchies and holons join
nested in the spiral
double coiled messages of life . . .

Stories in a book
recording the limitations of growth
preparing the way . . .
cultivating and nurturing memories
      remembered
in the petals of the rose
in the bounty of the lotus

Peirce and Spencer-Brown: History and Synergies in Cybersemiotics

Louis H. Kauffman and Søren Brier

In this issue of Cybernetics and Human Knowing we have a collection of papers devoted to the cybernetics and mathematics of Charles Sanders Peirce with a special focus on its synergies with Spencer Brown's thinking. We hope that the theme in this issue reflects the extraordinary richness of C. S. Peirce's work and his relevance to our present concerns and creativity in cybernetics. The theme is part of the journal's development of the area of cybersemiotics where Peirce and Spencer-Brown each have delivered the logical foundations. The similarities in the focus on some of the deep foundational subjects are astonishing, amongst those especially the concept of the void or Firstness and the continuum plus the continuity of mind and matter!

Peirce was a truly original American philosopher and logician working in the late 1800's and early 1900's on the east coast of the United States. His father, Benjamin Peirce was a mathematician at Harvard University, a pioneer in abstract algebra and well known for the dictum "Mathematics is the Science that draws necessary conclusions." Peirce was brought up in the intellectual environs of his father, and carried the philosophy of mathematics and logic to heights unimagined. Peirce, for personal and political reasons, spent the last 20 years of his life in virtual isolation from teaching, and in this time wrote volumes of material on mathematics, logic and philosophy that, thanks to his wife and friends, have been preserved and published after his death. These works are a continuing source of ideas and inspiration to scholars of mathematics, logic and cybernetics.

The papers in the thematic part of this issue are each reflections on Charles Sanders Peirce's work, its historical and present influence. The article "One, Two, Three ... Continuity: C. S. Peirce and the Nature of the Continuum" by Robin Robertson is a very accessible introduction to Peirce's revolutionary views on infinity and the continuum that goes beyond the present basis for mainstream mathematics conception of infinitesimals. Peirce's view is compared with Zeno's, Weirstrass, Dedekind and Cantor's and the problems their solutions generate. Robertson points out how Peirce's Synechism connects his metaphysics and his view on mathematics and meaning into a unique viewpoint that may be the foundations from which present, seemingly insurmountable, problems in mathematics and science can be reformulated in a fruitful way connecting to Spencer-Brown and Bohr's work, dealing with the limits of mechanicism. The article "Precursors to Laws of Form in C.S. Peirce's Collected Papers" by Jack Engstrom shows how Peirce's mathematics is related to the distinction based mathematics and philosophy of G. Spencer-Brown. He further discusses the metaphysical conceptions of the void or the unmarked state relating it to the mystical worldview. "In the world's most complicated card trick" Robertson tells the story about Peirce's combination of mathematics and a magical card trick that could have made him a fortune. The article "Peirce's Influence on Today's Mathematical Logic" by William Howard discusses Peirce's influence on modern mathematical logic and second order cybernetics from a historical perspective. He points out Peirce's influence on especially the semantic aspect of mathematical logic. The article "The Mathematics of Charles Sanders Peirce" by Louis H. Kauffman discusses the structure of the diagrammatic systems that Peirce used for logic and for mathematics. He starts with an analysis of Peirce's "Sign of illiation", which is a "portmanteau" sign (a sign containing two or more functions). The article shows the direct connection of these systems with the corresponding system of Spencer-Brown. It further shows the way in which the evolution of the diagrammatic systems is related to the evolution of logic, philosophy, cybernetics and mathematics such as the evolution of a Sign of itself (Peirce's view of the human condition). It ends on reflective note "We ourselves are portmanteau signs of a complex order. We are packing cases of multiple meanings large enough to make a human being a sign of itself." Finally this article explicates the relation of the amazing structure of the logical garnet of Shea Zellweger (a construction well known to Peirce scholars) to the work of both C.S. Peirce and Spencer-Brown. The article "Signs in action: Tarot as a self-organized system" by Inna Semetsky analyzes the structure of the Tarot and the relation of symbols to the Jungian psychology of the unconscious in Peircian terms.

The ASC column "Cybernetics of Fixed Points" by Louis H. Kauffman is a fictional dialogue in a bar among imaginary representatives for cybernetics (von Foerster, Maturana) and logic (Russell, Church) representing views of both dead and present researchers. The universe is the bartender and Peirce is one of the visitors.

The column, *A (cybernetic) musing: Constructing my cybernetic world*, by Ranulph Glanville draws a process oriented self-reflecting line through more than 30 years of his own work in cybernetics where "he is not trying to do science or anything like that: "I am trying to set up a system within which we can have such a science." He then very clearly describes the his project as "..how can we account for a world which each of us sees differently, and which, as a result, we cannot be sure is the same world?", and summarizes the basic results of this endeavor so far.

The book reviews are by Nina Ort of the new German edition of *Handbuch der Semiotic* by Winfied Nöth, which is one of the few semiotics handbooks that take small detours into information theory, cybernetics and cybersemiotics, comparing viewpoints making comparisons of viewpoints. Winfried Nöth reviews *Semiotica's* special issue *Biosemiotica* edited by Jesper Hoffmeyer and Claus Emmeche. This book is one of the newest and most comprehensive discussions of the enlargement of semiotics from human language to encompass the sign games of all living systems, thereby covering a great part of the same area as cybernetics and information theory, which also inspire several articles.

The artist of this issue is Jørn Særker Sørensen. Poems by Lawren Bale.

We welcome Robin Robertson as a new consultant editor.

Untangling Organic Koans

Crickets and cicadas sound
August with open windows
Squirrels barking   summer's end
Adrift, as we are, in this sea of chaos
    At the boundaries of each specific frame
We can taste and consume the messages
    Recycle their colors as sound
        And fashion the harmonics of love

        At the boundaries of each fractal
    Nothing appears to be moving
Yet I share with you these discrete items of infinite variety
    Increased complexity, elegance & ordered simplicity

Within this specific frame
    At the boundaries of our six senses
        Love and compassion forge life
    And we share with all the living this sense of awe
In the shared recognition of order, emergent pattern
    Ratio and degree

Crickets and cicadas sing
August   with open windows
Squirrels barking   summer's end

One, Two, Three . . . Continuity: C.S. Peirce and the Nature of the Continuum

Robin Robertson[1]

Abstract: The nature of the one and the many is an immemorial problem. This paper begins with Parmenides and the paradoxes presented by his disciple Zeno, then presents at some depth the mathematical concepts of *limit* and the *continuum* developed in the second half of the 19th-century by Karl Weierstrass, Richard Dedekind, and Georg Cantor. These interpretations are explicitly contrasted with C. S. Peirce's view of the nature of the continuum, and how this implies the actual existence of *infinitesimals*. A brief description of Peirce's concept of "one, two, three" is presented, showing how this new view on continuity completes this model of all reality, which he now termed *synechism*. Finally, several modern scientific examples of a similar view of continuity are presented.

Infinity is nothing but a peculiar twist given to generality.
 – Peirce in a letter to William James, June 8, 1903, (CP8:268).[2]

Twenty-five hundred years ago, Parmenides of Elea argued that all is one, that there is a single underlying principle of being. In Plato's *Parmenides*, we hear Parmenides gently draw out the ideas of a young Socrates. Then, at much greater length, he guides Aristotles (sic) ineluctably to a final conclusion: "whether one is or is not, one and the others in relation to themselves and one another, all of them, in every way, are and are not, and appear to be and appear not to be" (Plato, nd).

In other words, as soon as we look at the relationship between the *one* and the *many*, we're in deep waters, where it's difficult to be sure of much of anything. Parmenides' disciple Zeno argued against "the many" using a series of brilliant paradoxical arguments that are as fresh today as when they were first coined. The best known is undoubtedly that of Achilles and the Tortoise. Despite the fact that Achilles was the fastest of humans, and the Tortoise the slowest of animals, if the Tortoise is given a head-start on Achilles, no matter how small, Achilles can never catch him. For, by the time Achilles has arrived at the point where the Tortoise started the race, the Tortoise has moved forward. By the time Achilles comes to that point, the Tortoise has again moved. And so on.

Since, of course, we know that in a real race, Achilles would catch and pass the Tortoise, there must be something wrong with our reasoning. And what could that

---

[1] General Editor, *Psychological Perspectives*; P.O. Box 7226, Alhambra, CA 91802-7226, USA. E-mail: rrobertson@pacbell.net.

[2] **CP** *Collected Papers of Charles Sanders Peirce*, Volumes 1-6 edited by Charles Hartshorne, Paul Weiss. Volumes 7-8 edited by Arthur Burks. Cambridge: Harvard University Press. References in text indicate volume and paragraph number (CP1.11)

be? Zeno (and we will see, Peirce) would argue that the problem is the assumption that time and space are infinitely divisible. *Instead, both would say* that space and time are each indivisible concepts, impossible to break into parts. As Peirce says: "All the arguments of Zeno depend upon supposing that a *continuum* has ultimate parts. But a *continuum* is precisely that every part of which has parts, in the same sense. Hence he makes out his contradictions only by making a self-contradictory supposition" (CP5.355).

Despite numerous attempts by Aristotle and later philosophers, such as the Scholastics, in essence, Zeno's arguments remained unanswered until the second half of the nineteenth century, when a new foundational base of mathematics, *set theory*, led to rigorously defined concepts of *infinity*, *limit*, and the *continuum*. Three creators of set theory were Karl Weierstrass, Richard Dedekind, and, especially, Georg Cantor. The mathematical edifice that they built has, despite cracks in its own foundation, lasted to this day.

Standing outside this group, fully aware of their developments, but looking in from his own unique perspective, was *Charles Sanders Peirce* [1839-1914]. His view was, like all his views, idiosyncratic. Whereas Weierstrass, Dedekind and Cantor saw the continuum as a construction of points, Peirce regarded it as an entity in itself, beyond any construction. His view of the continuum offered the possibility for a continuity throughout matter and mind, living and dead, all supposed dichotomies. Thus he offered not only an alternative view of the mathematical continuum, but a different view of the *one*, which provides a possibility of finally advancing past Parmenides. I beg the reader's indulgence while I first present, at some length, the traditional view of the continuum, as developed in set theory, then contrast it with Peirce's view.

Weierstrass' Attempt to Define Infinite Limits in Finite Terms

> It is essentially a merit of the scientific activity of Weierstrass that there exists at present in analysis full agreement and certainty concerning the course of such types of reasoning which are based on the concept or irrational number and of limit in general.
> – David Hilbert (Struik, 1948).

If mathematics only dealt with finite numbers, there would have been no need for set theory. There would also have been no calculus. When Newton and Leibniz jointly created *calculus* late in the seventeenth century, they provided mathematicians and scientists with the most powerful mathematical tool ever created. Calculus enabled scientists to quantify both their observations and their theories to an extent hitherto inconceivable. But calculus depended on infinite and infinitesimal processes, and mathematicians of the time had only the vaguest of ideas what infinite and infinitesimal processes might actually mean. During the 18th century, mathematicians were largely satisfied to extend calculus, without questioning its foundation. As the 18th century gave way to the 19th century, mathematicians began to question that blithe attitude. By the second half of that century, they were ready to attack the issue head-on. In order to understand the approach they took, we need to begin with a short summary of what calculus is and does.

Imagine an irregular closed figure drawn on a piece of paper. Purely as an example, picture something like a lumpy circle. How can we find the area occupied by that shape? We can start by drawing a rectangle that is as small as possible yet fully contains the figure. It's simple enough to determine the rectangle's area, which is the product of its length times its height. In mathematical terms, we can call the area of the rectangle an *upper limit* on the area of the figure; that is, by using the area of the rectangle as an approximation to the area of the lumpy circle, we can insure that the true area must be smaller than our estimate.

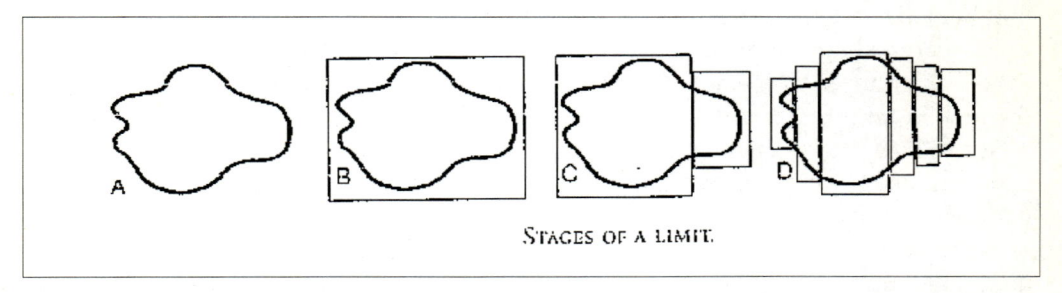

STAGES OF A LIMIT.

If, like our lumpy circle, the figure is quite asymmetric, we can improve our estimate by drawing two rectangles of different sizes next to each other, which together cover the whole figure. By calculating the area of each rectangle and adding the two, the upper limit will be smaller than before, closer to the actual area of the figure. If we increase the number of rectangles, our estimate will get better and better. The limit will get smaller and smaller, and approach the actual area more and more closely. With a thousand rectangles, the limit would be so close to the actual area that for all practical purposes, we could consider it to be identical. However, it's important to realize that it would still not be exact.

Now make a leap in thought! Imagine that we extend the number of rectangles endlessly. If we had some way of calculating the limit of this process by adding up an infinite number of such areas, the limit would no longer be merely an approximation to the area of the figure; it would be exactly the area of the figure. That is just what calculus does. Despite the fact that Newton and Leibniz were able to show how to calculate such limits, no one knew exactly what they were.

In the mid-19th century, German mathematician Karl Weierstrass [1815–1907] asked what do we really mean when we say that the sum approaches a limit? Simply that the sum of the rectangles can be made to differ from the limit by as small a quantity as we desire. More explicitly, if any desired difference is named, it is possible to pick a sufficient number of rectangles that their difference between their areas and the limit will be less than that quantity.

For example, say the limit is calculated to be 25 square inches using calculus. Then pick some very tiny number for the desired difference: say one thousandth of a square inch. Weierstrass says that we can find a larger enough number of rectangles that their sum will be within one thousandth of a square inch of 25 square inches. We'll say, just an example, that it takes 250 rectangles to do this. If we decide to pick a still smaller difference, say one millionth of a square inch, perhaps it might take 5,000 rectangles. But no matter how small the difference desired, it is possible to come up with a sufficient number of rectangles such that the sum of their areas differs from the limit by less than that difference.

The significance of Weierstrass' method is that he was able to rigorously define limits without ever mentioning infinity!

Weierstrass then used an extension of the same technique to discuss irrational numbers. As early as the 6th century B.C., Pythagoras had discovered that SR(2) (i.e., the square root of 2) could not be expressed as the ratio of two counting numbers (i.e., 1, 2, 3, ...). Over time, mathematicians had come to realize that there were a huge number of such numbers, which they called irrational—not a ratio—in contrast to so-called *rational* numbers like ½, or 23/47, ...; i.e., fractions formed by taking the ratio of two whole numbers. Though irrational simply meant "not rational," and hence had no pejorative meaning, it was also true that the term was singularly appropriate because of the degree of unease it caused Greek mathematicians.

Since irrational numbers could not be expressed as a simple ratio of integers, mathematicians were forced to use mathematically complex equations involving infinite series in order to calculate their value. This meant that they presented the same sort of problems involving infinity as calculus. Weierstrass got around this problem using a similar technique to the one that was so successful in dealing with limits. He defined an irrational number using a set containing a sequence of rational numbers that approached the actual value of the irrational number as a limit.

For example, the SR(2) which caused Pythagoras so much trouble, can be expressed in decimal notation as 1.414213... (Here our 3 little dots "..." have to do double duty, and mean not only that there is no end to the decimal expansion, but also that neither is there any definable pattern to it.) Weierstrass would define the square root of 2 as the set {1.4, 1.41, 1.414, 1.4142, 1.41421, 1.414213, ...}.[3] If we pick some tiny difference as we did with limits—say $1/10^4$ (i.e., one ten-thousandth)—we know that the 4th member of the set (i.e. 1.4142) is less than $1/10^4$ from the actual value of the SR(2). Similarly for a desired difference less than $1/10^6$ (i.e., one millionth), we have only to move to the 6th member of the set.

Once again, as with limits in calculus, the need to deal with infinity is avoided and Weierstrass manages to deal only with finite numbers. It is important to stress that Weierstrass didn't say that this set—{1.4, 1.41, 1.414, 1.4142, 1.41421, 1.414213, ...}—*approached* the irrational number as a limit (that is, "at infinity"); what he actually said was that the set itself *was* the irrational number. Though this

---

[3] Throughout the rest of this paper, I'll {}'s to indicate sets.

might seem an insignificant difference, he was shifting to an important new frame of reference: *set theory*.

What is a set? A set is merely a collection of things of the same kind; e.g., the set of books about logic is composed of all books about logic; the set of proofs for the existence of God is composed of all such proofs. Each item of the given kind is called a member of the set. It's important to grasp that the set is not the same as the members—the set is the collection, the assemblage, not the things assembled. Weierstrass was considering sets of numbers, but sets are not limited to numbers. Sets can include anything whatsoever.

Weierstrass' definition of the irrational numbers was important because it shifted the emphasis from an infinite series to a set where an irrational number could be defined to any degree of precision, with only a finite number of members of the set. However, clever as this technique was, many mathematicians objected to it as a trick, since any set which defined an irrational number still had an infinite number of members, regardless of the fact that Weierstrass was able to limit his discussion to at most a finite number of the set's members. These mathematicians were never to be satisfied by any subsequent improvement on Weierstrass' technique. The shift to set theory, however, opened the door to a new, exciting world for most mathematicians, and prepared the way for still more clever attempts at describing the number line. Drawing on the concepts of set theory, both Georg Cantor and Richard Dedekind (who we should note were contemporaries with C. S. Peirce) developed new definitions of irrational numbers. Dedekind's technique, called the *Dedekind Cut*, has become traditional.

The Dedekind Cut

> Dedekind came to the conclusion that the essence of the continuity of a line segment is not due to a vague hang-togetherness, but to an exactly opposite property—the nature of the division of the segment into two parts by a point on the segment (Boyer, 1968, p. 607).

Dedekind struggled with the same sort of problem addressed by Weierstrass, but focused on a different concept: the mathematical continuum. In mathematics, the continuum is another name for the number line; the unbroken line we are used to seeing, which has "0" in the middle and extends out indefinitely to the right with positive numbers, to the left with negative numbers.

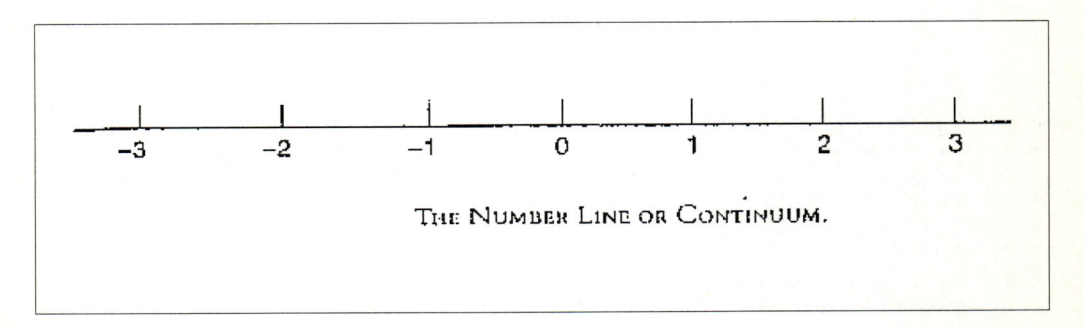

THE NUMBER LINE OR CONTINUUM.

But, though we draw numbers on it, the continuum itself is a line; hence a geometric concept, involving space. For example, Zeno's paradox of Achilles and the Tortoise can be pictured on the continuum. Alternately, the view of the mathematical set theorists like Weierstrass, Dedekind and Cantor (but, we will see, not C. S. Peirce, as we see below), is that the continuum is also the set of all *real* numbers (i.e., both rational and irrational numbers), and is, hence, also an arithmetic concept. *In this view, every spot on the number line corresponds to either a rational or irrational number; going in the other direction, every rational or irrational number has its place on the number line.*[4]

Rather than the whole continuum, let's simply take a chunk of it, say the part between 0 and 1 (any part will do as we'll see later in discussing Cantor's theory of transfinite numbers). According to the view of Weierstrass, Dedekind, and Cantor, every place on that line segment is either a rational or an irrational number. As you'll recall, the rational numbers are the numbers that represent fractions; hence each has a unique decimal identity. In between lie the irrational numbers, which cannot be expressed by an patterned decimal expansion, either finite or infinite. That's the problem with identifying irrational numbers. We showed Weierstrass' solution where he defined each irrational number to any degree of accuracy by a finite set of rational numbers. In contrast, Dedekind proposed a solution based in a mind experiment (Dedekind, 1888).[5]

Dedekind asked us to imagine a blade with an infinitely thin blade (i.e., it has no thickness at all) which can be used to cut the continuum into two segments. Since every point on the line corresponds to either a rational or an irrational number, the blade has to have hit a point corresponding to a number. And since the blade has no thickness at all, that point has to end up in either the left or the right segment. Since it doesn't matter, we'll assume that it always ends up in the leftmost segment. Since all the points in the left segment precede all the points in the right segment, every number in the left segment has to be less than every number in the right segment. So every number — rational or irrational — can be defined by cutting the number line into two segments in this way. And every such cut — which is traditionally called a *Dedekind Cut* — corresponds to a unique number.

Of course, it is implicit in Dedekind's solution that every point on the line interval is either a rational or an irrational number. As we'll see, Peirce strongly disagreed with this assumption.

---

[4] The attempt to find a way to equate number and geometry is as old as mathematics, and has led to both outstanding advances in mathematics, and to deep philosophical problems. For example, in the 17th century, Rene Descartes developed analytic geometry, which translated geometric positions and shapes into numeric coordinates and algebraic equations. Early in the 19th century, the greatest mathematician of all time, Karl Friedrich Gauss, went the other direction and translated *imaginary numbers* (which for lack of space we won't address in this paper) into geometric positions.

[5] Interestingly, Dedekind had been anticipated in this insight by the Greek thinker Eudoxus (c. 390 B.C.), as incorporated into Book V of Euclid's *Elements* (Kramer, 1982, p. 35).

Peirce's Approach To The Dedekind Cut

> ... my notion of the essential character of a perfect continuum is the absolute generality with which two rules hold good, first, that every part has parts; and second that every sufficiently small part has the same mode of immediate connection with others as every other has.
> (CP4.642)

Though Peirce never directly discussed the Dedekind Cut, he was aware of his work and did express a view of the continuum that was in marked contrast. In this section, I'm going to use Peirce's view of the continuum to produce a *reductio absurdum* argument against the Dedekind Cut.

Clearly, both Weierstrass and Dedekind assume that the continuum, as represented by a line, is composed of individual points (each point of which can be expressed by either a rational or an irrational number). Therefore, in Peirce's view, since there are no holes in the line, and since the line is supposedly made up of points, even before we cut the line in two, we must be able to exactly identify the cutting-point. That is a clear consequence of each point having an individual identity.

But then, because the blade has no width at all, it must be possible to divide the cutting-point into two points, one in the left region {A}, and one in the right {B}. We can do this, because, in Peirce's view, any "point" on the continuum must be like a raindrop which the cut divides into "two ideal rain drops, distinct but not different"[6] (CP4.311). Dedekind ends up with the point in either {A} or {B} (we chose {A} for simplicity) because he views points as the "atoms" of the line. *Peirce insists that there are no "atoms," only "parts," which can be divided endlessly.*

After the cut, in Peirce's view, we can now see an ordering that wasn't there before; that is, the rightmost point of {A} has to be earlier in sequence (i.e., to the left) than the leftmost point of {B}. But, on the other hand, if we rejoin the two parts, the two points have recombined again into a single point. Obviously an impossibility: a point can't both be an indivisible "atom" and also capable of being divided into further "atoms." Hence, the line is not made up out of points; it is only the act of making the cut which creates an ordering of points of a line. Or put another way, *points don't exist except when we create them.*

In Peirce's words: "the points on a line not yet actually determined are mere potentialities" (CP3.568). *"Mere potentialities."* Hence, for Peirce, the only way to create points is to identify them by some act of distinction. Yet, no matter how often we create points by making such distinctions, we can never exhaust the points on a line, since there are always unlimited sequences of points between any two points we identify.

If there is room on a line for any multitude of points, however great, a genuine continuity implies, then, that the aggregate of points of a line is too great to form a collection: the points lose their identity; or rather, they never had any numerical

---

[6] A phrase Peirce used with respect to permutations, but which fits perfectly here.

identity, for the reason that they are only possibilities, and therefore are essentially general. They only become individual when they are separately marked on a line; and however many be separately marked, there is room to mark more in any multitude. (CP7.209)

How Distinctions Create Worlds

> . . . when one element cannot even be supposed without another, they may ofttimes be distinguished from one another. Thus we can neither imagine nor suppose a taller without a shorter, yet we can distinguish the taller from the shorter. I call this mode of separation *distinction*. (CP1.353).

In 1903, trying to find a way to express the nature of true continuity, Peirce recorded the following parenthetical thought in his personal copy of the Dictionary.[7]

> It seems to me to point to this: that it is impossible to get the idea of continuity without two dimensions. An oval line is continuous, because it is impossible to pass from the inside to the outside without passing a point of the curve (CP6.165).

This is a perfect expression of what Peirce means by points being "mere potentialities" until brought into existence by an act of distinction. In his example, in order to bring a point into existence, one has to pass from one side of a Spencer-Brown distinction to another. In a previous article (Robertson, 1999), I described how, in his *Laws of Form* (1979), logician G. Spencer-Brown presented a mathematical system which dealt with the emergence of anything out of the void. *Laws of Form* traces how a single *distinction* (a "mark") in a void leads to the creation of *space*, where space is considered at its most primitive, without dimension. This in turn leads to two seemingly self-evident "laws." With those laws as axioms, first an arithmetic is developed, then an algebra based on the arithmetic, which is formally equivalent to Boolean algebra.

When you make a Spencer-Brown distinction, there are now two things: the "mark" and the "not-mark." You've created a "space" and that's all that there is in the space you have created. The mark and the not-mark form a necessary and complementary pair without which space—any space—cannot exist. But, once you use his two "laws" to create an arithmetic, you are able to string marks together to form a series of signs that appear to be something other than the mark and the not-mark. Until you apply the two laws, these series of signs are indeterminate: you don't know whether they reduce to the mark or the not-mark. They haven't yet "collapsed" into the mark or the non-mark.

Note how similar the views of Peirce and Spencer-Brown are to the Copenhagen model of quantum mechanics. For Peirce, points and their distinctions are created by the process of marking or referencing them, and their

---

[7] Now in the Treasure Room at the Widener Library, Cambridge, Mass.

existence and distinctions collapse when we change our frame of reference. In the Copenhagen model, reality is indeterminate until, under observation, it appears as either a particle or a wave. When observation ends, the particle or wave once again collapses into indeterminacy. In his biography of Werner Heisenberg, author David C. Cassidy discusses how, when Heisenberg was first introducing his famed "uncertainty principle," basing it on a particle view of reality, Niels Bohr argued with his younger colleague that he should instead view it within the "complementarity principle," in which both waves and particles are necessary, complementary aspects of physical reality.

That is, we know particles have energy (E) and momentum (p). Similarly we know waves can be described fully in terms of space (q) and time (t). Furthermore, we know that all causal descriptions are based in one or more of these four variables . However, each uncertainty combines possibilities of both waves and particles. In fact, it's because we combine two incompatible elements—particle and wave—into a single possibility, that we have an inherent uncertainty. Bohr thus argued that reality emerges from uncertainty by the act of observation, not by the process of discovery. "Heisenberg rose from his seat in the audience that day to confer his public approval on Bohr's interpretation of their physics. The Copenhagen interpretation was born" (Cassidy, 1992, 244).

Thus Heisenberg and Bohr, dealing with the physical world at the quantum level, and Spencer-Brown and Peirce, dealing with the most primitive levels of mathematics and logic, arrived at a point where the act of distinction creates reality out of a previously undetermined state.

Cantor's Theory Of Transfinite Numbers

> Both in his time and in the years since, Cantor's name has signified both controversy and schism. Ultimately, transfinite set theory has served to divide mathematicians into distant camps determined largely by their irreconcilable views of the nature of mathematics in general and of the status of the infinite in particular.[8]

Questions about the continuum and the nature of points raise issues about the nature of infinity. The third, and greatest of our trio of mathematicians who developed set theory was Georg Cantor. Cantor's theory of transfinite numbers was a major influence on Peirce, and, hence, needs to be discussed before we move further into Peirce's own approach.

Long before Cantor, mathematicians realized that there is no end to the integers; if a largest integer could be conceived, merely add one to it and it's no longer the largest. There is also clearly no end to the points on a line provided that a point is understood to be without dimension. Cantor had the brilliance to ask the seemingly silly question: Are these infinities both the same size. Lest this sound like the medieval scholastic arguments about the number of angels who could

---

[8] Joseph Warren Dauben, *Georg Cantor: His Mathematics and Philosophy of the Infinite*, p. 1.

stand on the head of a pin, the reader should be aware that Cantor found a way of quantifying infinity, and thus answering his own question.

Let's return to the concept of a set. What does it mean to say that two sets have the same number of members? For example, what does it mean to say that the set of fingers has the same number of members as the set of toes? If we think deeply about it, it merely means that we can pair off members of each of the two sets and have no members left over in either set. In our example, we could place each of our ten fingers on one of our ten toes.

Notice that it wouldn't matter what finger we match with what toe as long as we were careful not to pair any finger with two toes or any toe with two fingers. Further, we could define a *cardinal number*, in this case 10, which would characterize the number of members in each set. Having developed the cardinal number 10, it could then be used to describe the number of members of any set which can be paired off with the set of fingers. We would describe each as having the same *cardinality*. This provides an unequivocal way to compare sets to see if they are the same size.

This method of counting by pairing the members of one set with the members of another is commonly termed the *pigeon-hole* technique. Cantor found that the pigeon-hole technique produced surprising results with infinite sets. For example, he discovered that there were exactly as many even integers as there were both odd and even integers. He reasoned that for every integer i in the set of all integers, he could match i with the integer $2i$ in the set of even integers. For example, *1* in the set of all integers would be matched with *2* in the set of even integers, *2* with *4*, *3* with *6*, etc. No matter what integer you named in either set, this method uniquely defines the integer it corresponds to in the other set. Thus the pigeon-hole technique proved that the set of even integers is the same size as the set of all integers. Try it if it doesn't seem reasonable. At first it will seem obvious that there are more integers than even integers, since the odd integers are left out. But that isn't the point. If the two sets can be matched without any members of either set being left over, then they have the same cardinality. The problem isn't with the method, it's with infinity itself.

Cantor could use the same logic to show that the set of numbers evenly divisible by three is also just as big as the set of all integers. In fact the particular multiple made no difference; the set of numbers evenly divisible by a billion is still just as big as the set of all integers. Going in the other direction produced the same result. The integers are a sub-set of the rational numbers (remember: the fractions.) For example, 2 can be expressed as 2/1 as a fraction; 3 as 3/1, etc. But think of all the other fractions: ½, 1/3, 357/962, etc. Surely there are more fractions than integers! Instead, Cantor used an elaboration of the pigeon-hole technique to prove that the set of rational numbers had exactly as many members as the set of integers. This was a very surprising result indeed! In fact, Cantor proved that *any infinite sub-set of the set of rational numbers contains exactly the same number of members*! All have the same cardinality, which Cantor termed

Aleph-0. This number is commonly referred to as *countably infinite*, since all such sets can be paired off with the counting numbers 1, 2, 3, ...

So we see that when we are dealing with infinite sets, it is possible to match the whole set with a sub-set of itself, with nothing being left out. Richard Dedekind has traditionally been credited with this lovely definition of infinite sets: "A system S is said to be *infinite* when it is similar to a proper part of itself, in the contrary case S is said to be a *finite* system" (Dedekind, 1888). But Peirce had been there before Dedekind, using the same distinction in his paper "The Logic of Number" (CP3.288). Peirce also said that he had communicated this idea in a letter to Dedekind. Peirce was especially proud of this discovery, claiming that "the proposition that finite and infinite collections are distinguished by the applicability to the former of the syllogism of transposed quantity ought to be regarded as the basal one of scientific arithmetic" (CP6.114). He produced many versions of this argument in his papers. Here's one:

> Every Texan kills a Texan,
> No Texan is killed by more than one Texan,
> Hence, every Texan is killed by a Texan (CP 3.288).

If there is a finite number of Texans, this is true, but if, on the other hand George W. Bush's dearest wish would happen and there were an infinite number of Texans, then this would not be true. For example, Texan #1 could kill Texan #2, Texan #2 could kill Texan #3, and so on. Texan #1 would remain alive. Or, if we want more living Texans, Texan #1 could kill Texan #2, Texan #2 could kill Texan #4, Texan #3 could kill Texan #6, and so on. An infinite set of Texans would still be alive at the end of this killing spree (i.e., Texans #1, 3, 5, ...).

Georg Cantor went beyond this and asked if every infinite set was countably infinite; i.e., is every infinite set the same size as every other infinite set? The answer to that was even more surprisingly a resounding no. Let's turn to his idea of power sets to see why there is more than one size to infinity; why there are, in fact, an unlimited number of infinities.

For any set, Cantor defined its power set as the set of all the sub-sets of the original set. Key here is that the power set is one level higher than the original set: its members are themselves sets. For example, let's take the set of primary colors: {red, yellow, and blue}. The power set would consist of all the sub-sets of this set; first taken one at a time, then two at a time, finally three at a time.[9] So the power set would be: **{**{red}, {yellow}, {blue}, {red, yellow}, {red, blue}, {yellow, blue}, and {red, yellow, blue}**}**.[10]

Another way of looking at a power set is to consider it as composed of all the relationships between the original members of a set, all the ways they can be

---

[9] Actually even taken zero at a time. A special set called the *null set* was needed to complete set theory, just as zero is needed to complete the number line. In all of our discussions of the continuum and real numbers, we have implicitly included zero.

[10] (I've used bold brackets **{}** to indicate that the power set is a set, and normal brackets {} to indicate that its members are sets.)

linked together in any combination. It is easy to see that as the original set gets bigger, the power set gets much, much bigger.[11] In our little example, the set of primary colors had 3 members, while the power set had 8 (including the null set as the 8th.) A set with 4 members would have a power set with 16 members, and so forth. Cantor was able to demonstrate that the power set is always of higher order of cardinality than the original set, *even if the set is infinite*. Thus the power set of the rational numbers has a cardinality bigger than *Aleph-0*, which Cantor termed *Aleph-1*. *Aleph-1* in turn has a power set *Aleph-2*, and so on. *So there is no biggest size for infinity, since we can always create a bigger infinity by taking the power set of the previous set.*

It is simple to demonstrate that the set of real numbers is equivalent to the first power set of the countable numbers. That is, the power set of the countable numbers is the set of all countable numbers taken one at a time, two at a time, and so forth: {1}, {2}, {3}, . . . , {1, 2}, {1,3}, . . . , {2, 3}, {2, 4}, . . . , {1, 2, 3}, . . . The real numbers can be expressed in decimal form as a decimal place followed by all the combinations of the real numbers taken one at a time, two at a time, and so forth: 0.1, 0.2, 0.3, . . . , 0.12, 0.13, . . . , 0.23, 0.24, . . . , 0.123, . . . So clearly the first power set and the real numbers are the same size.

Having shown that power set of the countable numbers was a larger size of infinity than the countable numbers themselves, and that endless, ever larger infinite sets could be constructed by taking power sets of power sets, Cantor proposed that this was all that there was to infinity. *This proposal is called Cantor's Continuum Hypotheses. Cantor's Continuum Hypothesis says simply that the power of the continuum = Aleph-1 (i.e., the power set of the countable numbers, which is equivalent to the set of all real numbers).*[12]

In famed mathematician Kurt Gödel's words:

> Cantor's continuum hypothesis is simply the question: How many points are there on a straight line in Euclidean space? An equivalent question is: How many different sets of integers do there exist? This question, of course, could arise only after the concept of number had been extended to infinite sets (Gödel, 1964, p. 254).

Cantor's hypothesis that the number of points is Aleph-1, which he was sure was correct, remains one of the great unresolved conjectures in mathematics. Again, quoting Gödel:

> But, although Cantor's set theory now has had a development of more than seventy years and the problem evidently is of great importance for it, *nothing has been proved so far about the question what the power of the continuum is*...Not even an upper bound, however large, can be assigned for the power of the continuum (Gödel, 1964, p. 256).

---

[11] The power set has 2^n members, where n is the size of the original set. Hence the notation 2^n is used to stand for the power set of any set of cardinality n (remember that cardinality is the same as size within Cantor's theory).

[12] There is also a generalized Continuum Hypothesis, but we won't address that here.

Thus Peirce's idea of the continuum as being too rich to be exhausted by the real numbers (or in fact, by any combination of numbers), seems less strange.

Infinitesimals & The Denial Of The Continuum Hypothesis

> I believe that adding up all that has been said one has good reason for suspecting that the role of the continuum problem in set theory will be to lead to the discovery of new axioms which will make it possible to disprove Cantor's conjecture (Gödel, 1964, p. 264).

Peirce was mightily impressed by Cantor's achievement and said so numerous times in his own papers; e.g., "in truth, there are but two grades of magnitude, the *endless* [by which Peirce meant countable infinity] and the *innumerable* [by which he meant the power sets]." In his early admiration for Cantor's theory, Peirce did not seem to feel that matching the continuum to the real numbers was impossible. In that earlier period of time[13], he still felt that "it is evident that there are as many points on a line or in an interval of time as there are real numbers, in all" (CP6.118).

It was only after struggling with the ideas of Weierstrass, Dedekind, and Cantor for nearly two decades that Peirce came to his own unique approach to the continuum. In a letter to Paul Carus, editor of *The Monist*, on Aug. 17, 1899, Peirce said that " . . . the true nature of continuity . . . is now quite clear to me." Previously, Peirce said, had been "dominated by Cantor's point of view." Now he saw that it was best not to try "to build up a continuum from points, as Cantor does" (Peirce, 1998b, p. xxii).

In Peirce's new view, the continuum is an inexhaustible source of numbers but does not, itself, consist of points or numbers! One way to visualize Peirce's argument is to imagine that first we approximate the continuum by the rational numbers; there will still be, however, "holes" between the rational numbers. We can then plug those hole with the irrational numbers.[14] But Peirce would say that even after adding the irrational numbers, there would still be holes. We can fill those holes with further sequences of numbers by taking the power sets of the irrationals. But, Peirce would insist that there will still be holes. So we can try filling those holes with still further sequences of numbers. According to Peirce, no matter how many sets of transfinite numbers we use, holes will still remain. It is in this way that we see how the continuum is considered to be inexhaustible and not just infinite. Though Peirce was probably the first to realize the basic problems with attempting to relate a spatial concept like the continuum, with any set of numbers, he wasn't alone in his belief.

---

[13] From roughly 1880 through the late 1890's.

[14] When I studied mathematics in college over thirty years ago, we took for granted that this filled all the holes. When I first encountered Cantor's Continuum Hypothesis, at first I couldn't even understand what it might mean. How could there possibly be any other numbers?

> As early as 1905 René Baire...suggested that Cantor's continuum hypothesis assumed the identifiability of two concepts that were intrinsically different and of noncomparable orders of magnitude...The two ideas were inherently antithetical: the nature of the continuum, regarded as the collection of all infinite sequences of integers was something totally different (Dauben, 1979, p. 269).

After producing his epochal Incompleteness Proof in 1931, Kurt Gödel spent much of his later mathematical life trying to resolve the Continuum Hypothesis. In 1940, he managed to prove that if a modified set theory, which does not include the Continuum Hypothesis, is consistent, then it will remain consistent if the Continuum Hypothesis is added as an additional axiom. In 1963, mathematician Paul Cohen was able to prove the reverse: If a modified set theory which does not include the Continuum Hypothesis, is consistent, then it will still be consistent if the continuum hypothesis is assumed to be false. In other words, between Gödel and Cohen, they had demonstrated that the Continuum Hypothesis was undecidable within set theory.

That being said, both Gödel and Cohen felt that ultimately, within some extension to set theory, the Continuum Hypothesis would be decidable, and both felt that it would be false. Gödel didn't go as far as Peirce, and at one time thought that perhaps the cardinality of the continuum might be Aleph-2. Dauben says that "in [Paul] Cohen's view, the continuum was clearly an incredibly rich set one produced by a bold new axiom which could never be approached by any piecemeal process of construction" (1979, p. 269). This view sounds close to Peirce, as both felt that it was impossible to arrive at the continuum by constructing it from points, or sequences of points. The continuum simply was, and points were simply abstractions for possibilities within it.

If this were true, however, then what number abstractions might we assign to "the holes in the continuum"? Well, if we go all the way back to the creation of calculus by Newton and Leibniz, we find that, in addition to limits, both theorists expressly assumed infinitely small particles (though both assured other scientists that these were merely intellectual abstractions). Newton called his particles fluxions, but they are better known as infinitesimals. Infinitesimals are greater than zero, yet less than any normal positive number. In a famous quote, 18th-century philosopher Bishop George Berkeley, lampooned fluxions:

> And what are these fluxions? The velocities of evanescent increments. And what are these same evanescent increments? They are neither finite quantities, nor quantities infinitely small, nor yet nothing. May we not call them the ghosts of departed quantities . . . ? (Smith, 1959, p. 633).

In contrast, Peirce came to regard infinitesimals as something very real, in fact, as the only way to explain the continuum: " . . . the infinitesimals must be actual real distances, and not mere mathematical conceptions like SR(-1)" (CP 3.570). In this assumption, Peirce was clearly going against the main-stream of late 19th and early 20th-century mathematics, which tried to avoid infinitesimals like the plague.

In our more recent time, two different mathematical approaches have emerged which explicitly admit infinitesimals into mathematics: the *surreal numbers* developed by the legendary John Horton Conway (Knuth, 1974; Conway & Guy, 1996)[15], and the *hyperreal* numbers that emerge from the late Abraham Robinson's *non-standard analysis* (Robinson, 1966).

Using only two simple generative rules, Conway is able to make all integers emerge, both positive and negative, then rational numbers, then the real numbers that fit between, then all of Cantor's transfinite numbers. And it just begins there. Infinitesimals, which are the inverse of the transfinite numbers, emerge, then algebraic roots of transfinites and infinitesimals, and so on endlessly.

But Abraham Robinson's non-standard analysis has had more impact on mathematics. Robinson took the set of real numbers **R**, then created an expanded set **R*** by adding all the infinitesimals and infinites. Infinites and infinitesimals are the inverse of each other (i.e., if x is infinitesimal, 1/x is infinite, and vice versa). Anything in **R***, not in **R**, is *non-standard*. All hyperreal numbers have a standard part and a non-standard part. I won't go into the details here, but what is special about Robinson's achievement is that anything proved within **R***, using non-standard analysis, is automatically true in **R**, using standard mathematical techniques. Peirce would have been fascinated not only with Robinson's achievement, but with the fact that non-standard analysis evolved out of logic, and needed one major theorem from logic.[16]

Peirce would have been less thrilled, however, with the fact that Robinson regards hyperreal numbers merely as notational tools, not as actually existent. But then Robinson was skeptical enough to have once remarked: "I am not sure I really BELIEVE in the set of all natural numbers! (Dauben, 1995, p. 354).

One, Two, Three . . . Continuity

> TRICHOTOMIC is the art of making three-fold divisions. Such division depends on the concept of 1st, 2nd, 3rd. First is the beginning, that which is fresh, original, spontaneous, free. Second is that which is determined, terminated, ended, correlative, object, necessitated, reacting. Third is the medium, becoming, developing, bringing about (Peirce, 1992, p. 280).

Throughout his writings, early and late, Peirce was fascinated by (one might almost say "obsessed by") the concept of "three-ness", "triads," "trichotomy," "One, Two, Three," etc. He even joked about this tendency. "I am a determined foe of no innocent number; I respect and esteem them all their several ways; but I am forced to confess to a leaning to the number three in philosophy" (CP1.355).

In "A Guess at the Riddle" (CP1.354-416), which the editors of *The Essential Peirce* consider Peirce's "greatest and most original contribution to speculative philosophy" (Peirce, 1998b, p. 245), Peirce attempts to extend this principle to all

---

[15] Most famed to the general public for creating the game of Life.

[16] Tthe compactness theorem.

reality, beginning with the essential concept, then extending it into reasoning, metaphysics, psychology, physiology, biology, physics, sociology (by which he means not the study of humans in groups, but the study of a "community of cells"), and theology. Perhaps the best-known of his hierarchies of triads proceeds from Quality — Relation — Representation (or Sign); then breaks Sign into Icon — Index — Symbol; then Symbol into Terms — Propositions — Arguments; and Arguments into Hypothesis (or Abduction) — Deduction — Induction.

I won't belabor these categories, as they are the core of Peirce's philosophy and logic, and will undoubtedly be treated at length by others. But there was a stumbling block implicit in the concept of "one, two, three": if the 3rd participates in both the 1st and the 2nd, it necessarily contains both within itself already. Or perhaps the 2nd and the 3rd are already contained within the 1st, and so forth. This is an inherent problem in any system which has to proceed from the ideal to the actual, via a hierarchical system. Both the Neoplatonists and the Gnostics struggled with this problem and evolved hierarchies which approximated the process. As Peirce expressed the issue in logic: " . . . were a proposition to be true up to a certain instant and thereafter to be false, at that instant it would be both true and false". (Eisele, 1976, vol. 3, p. xvii).

As Murphey expresses Peirce's problem, and its resolution through his late views on the continuum:

> [Peirce] required a property characterizing unactualized possibilities which would be itself actual so that it could be observed. Yet incredibly enough Peirce found such a property in 1896 in continuity. For by his definition of the continuum—and it must be borne in mind that he regarded his definition as the only one which avoided the paradoxes of set theory—any true continuum must contain potentialities which are not only not now actualized but which are greater in multitude than any set of events which can ever be actualized (Murphey, 1993, p. 395).

Since, for Peirce, the continuum contained all numbers that could ever be, *in potentia*, not in actuality, it was the perfect model for his triadic principle throughout all reality. Peirce contrasted *materialism* ("the doctrine that matter is everything"), *idealism* ("the doctrine that ideas are everything"), with *synechism* ("the tendency to regard everything as continuous")[17] (CP7.565). He made this quite explicit in this description of how true continuity is possible only within a triadic relationship:

> A potential collection, more multitudinous than any collection of distinct individuals can be, cannot be entirely vague. For the potentiality supposes that the individuals are determinable in every multitude. That is, they are determinable as distinct. But there cannot be a distinctive quality for each individual; for these qualities would form a collection too multitudinous for them to remain distinct. It must therefore be by means of relations that the individuals are distinguishable from one another. . . . No perfect continuum can be defined by a [asymmetrical] dyadic relation [since the origin and terminus would be points of discontinuity]. But if we take instead a triadic relation, and say A is r to B for C, say, to fix our ideas, that proceeding from A

---

[17] Peirce was always one to coin a new term, which often makes reading him a process of translation.

in a particular way, say to the right, you reach B before C, it is quite evident that a continuum will result like a self-returning line with no discontinuity whatever.... (CP6.188).

With synechism, everything flowed into everything else, through an ineluctable process of 1–2–3. With the concept of the continuum underlying synechism, he could confidently regard space and time as seamless. "It would be in the general spirit of synechism to hold that time ought to be supposed truly continuous" (CP6.170). With synechism, Peirce felt that he had answered Parmenides: "There is a famous saying of Parmenides, . . . 'being is, and not-being is nothing.' This sounds plausible, yet synechism flatly denies it, declaring that being is a matter of more or less, so as to merge insensibily into nothing" (CP7.569).

Unfortunately Peirce died before he could see the emergence of *quantum mechanics* and *chaos theory*. In the Copenhagen model of quantum mechanics, reality is neither a wave nor a particle, until actualized through observation. In Schrödinger's wave model of quantum mechanics, reality exists *in potentia* until the quantum wave collapses into physical reality. In David Bohm's model, there is an underlying wholeness, much like Peirce's continuum, in which all that can ever be exists *infolded in potentia*, until it is *unfolded* into what we know as the world (Bohm, 1980).

More recently, chaos theory has shown how throughout nature, order can give way to deterministic chaos, from which new order emerges. Illya Prigogine received the Nobel prize for his discovery that, as long as complex systems (*dissipative structures*) can bring in matter and energy from outside themselves, they can go through a period of instability, then emerge with a new self-organization (Prigogine, 1984).

Perhaps we can give Peirce the last word on the importance of his interpretation of the nature of the continuum, and the applicability of continuity throughout nature:

> I was led at the very outset to think that one great desideratum in all theorizing was to make fuller use of the principle of continuity. My attention was from the beginning drawn to the need of looking at matters in the light of that conception, but I did not, at first suppose, that it was, as I gradually came to find it, the master-key of philosophy (Parker, 1998, xiv).

Acknowledgments

I would like to express my grateful appreciation to Robert J. Porter, Ph.D. for his careful reading and editing suggestions to successive stages of this paper. Bob is the immediate past president of the Society for Chaos Theory in Psychology and the Life Sciences.

References

Bell, E.T. (1937/1965). *Men of mathematics*. New York: Simon and Schuster.
Bohm, David. (1980). *Wholeness and the implicate order*. Boston: Routledge and Kegan Paul.

Boyer, Carl B. *A history of mathematics*. Princeton: Princeton University Press, 1968.
Cassidy, David C. (1992). *Uncertainty: The life and science of Werner Heisenberg*. New York: W. H. Freeman and Co.
Conway, John H., & Guy, Richard K. (1996). *The Book of numbers*. New York: Springer-Verlag.
Dauben, Joseph Warren. (1977). C. S. Peirce's philosophy of infinite sets. *Mathematics Magazine*, 50:3, 123-135.
Dauben, Joseph Warren. (1979). *Georg Cantor: His mathematics and philosophy of the infinite*. Princeton: Princeton University Press.
Dauben, Joseph Warren. (1995). *Abraham Robinson*. Princeton: Princeton University Press.
Dedekind, Richard. (1888), *Was sind und was sollen die Zahlen*, W. W. Beman, trans., The nature and meaning of numbers, *Essays on the theory of numbers*. Reprint of edition of 1901. New York: Dover, 1963, pp. 6-115.
Eisele, Carolyn. (ed.). (1976). *The new elements of mathematics*. 3 vols. The Hague: Mouton de Gruyter.
Gödel, Kurt. (1964). What is Cantor's continuum problem?. In Solomon Feferman, editor-in-chief, *Kurt Gödel, collected works, volume II: Publications 1938-1974*. New York: Oxford University Press, 1990, pp. 254-270.
Knuth, D. E. (1974). *Surreal numbers*. Reading, MA: Addison-Wesley.
Kramer, Edna E. (1982). *The nature and growth of modern mathematics*. Corrected edition. Princeton: Princeton University Press.
Murphey, Murray G. (1993). *The development of Peirce's philosophy*. Indianapolis: Hackett Publishing Co.
Parker, Kelly A. (1998). *The continuity of Peirce's thought*. Nashville: Vanderbilt University Press.
Peirce, C. S. (1992). *The essential Peirce, vol. 1*. Nathan Houser and Christian Kloesel, eds. Bloomington: Indiana University Press.
Peirce, C. S. (1998a). *Collected papers of Charles Sanders Peirce*, volumes 1-6 edited by Charles Hartshorne, Paul Weiss. Volumes 7-8 edited by Arthur Burks. Cambridge: Harvard University Press.
Peirce, C. S. (1998b). *The essential Peirce, vol. 2*. The Peirce Edition Project, eds. Bloomington: Indiana University Press.
Peirce, C. S. (1998c). *The essential Writings*. Edward C. Moore, ed. New York: Prometheus Books.
Plato. *Parmenides*. (nd). Benjamin Jowett, Trans. From the World Wide Web:
Prigogine, Ilya & Stengers, Isabelle. (1984). *Order out of chaos: Man's new dialogue with nature*. New York: Bantam Books.
Robertson, Robin. (1999). Something from nothing: G. Spencer-Brown's laws of form. *Cybernetics & Human Knowing*, Vol. 6, no. 4, pp. 43-55.
Robinson, Abraham. *Non-standard analysis*. Amsterdam: North-Holland Publishing Company, 1966.
Smith, David Eugene. (1959). *A source book in mathematics*. Reprint of 1929 edition. New York: Dover.
Spencer-Brown, G.. (1979). *Laws of form* (rev. ed.). New York: E. P. Dutton.
Struik, Dirk J. (1948). *A concise history of mathematics*. (4th revised ed.). New York: Dover Publications, reprint 1987.

C. S. Peirce's Precursors to *Laws of Form*

Jack Engstrom[1]

Abstract: What Charles Sanders Peirce (1839-1914) called "the alpha part" of his "existential graphs" for logic quite precisely prefigures a number of features of G. Spencer-Brown's notational system in his book *Laws of Form* (1969). These two systems are compared and contrasted with each other and also with Euler-Venn diagrams. The significance of *Laws of Form* to other disciplines, in particular epistemology and the possibility of a metaphysical and even mystical understanding of knowledge, is discussed. Also, a modification is proposed of one of Peirce's typographical notations to give it *Laws of Form*-like properties.

Abbreviations: *Laws of Form* is abbreviated as LoF. References are indicated by square brackets: for example "[LoF]", and are collected in the References. If no author or title is indicated in the brackets, then it refers to Peirce's *Collected Papers*, where for example, [3.475] means Volume 3, paragraph number 475 (*not* page number).

1. Introduction

Charles Sanders Peirce (1839-1914) made original contributions to logic, corresponded with eminent logicians of his day, and influenced (*via* Schröder and Peano) modern logic and its notation [Encycl.Brit.]. In his earlier years Peirce created variations on notation for logic of the "linear" or typographical sort that has since become standard. But in his later years he conceived and elaborated a logical notation he called "existential graphs" (see later sections of this paper) that was more diagrammatic and iconic than typographic. He dubbed this work his *chef d'oeuvre* (masterpiece) to indicate how highly he valued it. Yet it "was little recognized until the 1960s." [Encycl.Brit.].

Coincidentally, George Spencer-Brown developed his book *Laws of Form* in the 1960s. Like Peirce, Spencer-Brown (in *Laws of Form*) was concerned with crafting a notation for the mathematical structure of logic, and created a notation that was diagrammatic and not simply typographic. Beginning his explorations in late 1959 [Spencer-Brown, from the Preface], he published in 1969. His exploration became a quest that arrived at the vision that the various forms encountered in our experience and existence arise in stages out of formlessness by drawing a distinction and then arranging indications or tokens of that distinction

---

[1] Maharishi University of Management SU-129, 1000 N. 4th St., Fairfield, IA 52557. E-mail: engstrom@kdsi.net.

[Engstrom 1999, pages 33-35, 39-40], [Spencer-Brown 1972: v].

As an aficionado (since 1973) of *Laws of Form* (hereafter abbreviated to "LoF"), I had heard that it was in some respects anticipated by Charles Sanders Peirce, particularly in his "entitative graphs" and "existential graphs" (see later sections of this paper). So I approached writing this paper as a welcome opportunity to acquaint myself with Peirce's *Collected Papers*, and to assess for myself any features shared with LoF. I wait until the final section of this paper to consider whether Peirce influenced Spencer-Brown in creating his LoF system.

Preview of sections
The main part of this paper offers instances in which Peirce's work, especially what he called the alpha part of his existential graphs, prefigured Spencer-Brown's LoF notation. But to set this rather technical paper into a larger perspective, we begin by discussing the significance of *Laws of Form* to other disciplines. This discussion is continued in two of the appendices.

Both Peirce's work and LoF provide notations for logic. Since the reader is not expected to be familiar with Peirce's work, LoF or logic, we give primers or vignettes of logic and of LoF in the two sections following the section on significance of LoF. In these primers we introduce the basic concepts, technical terms, and symbols that will be used later on in the text.

All this done, we at last begin listing the specific features of LoF notation to be correlated with Peirce's work, and proceed in three sections and their various subsections to give the correlations. In the process, relevant parts of Peirce's existential graphs and other work are explicated. After showing all of these similarities, the next section points out three differences between the two systems. Another section then introduces Venn diagrams, and yet another compares and contrasts the three systems with each other.

In the final section we examine to what extent Spencer-Brown was familiar with, and LoF influenced by, Peirce. In an appendix, I propose a modification of Peirce's use of a horizontal line to give it LoF-like properties.

2. *Laws of Form's* significance

To set this rather technical paper into a larger perspective, we will now discuss the significance of *Laws of Form* to other disciplines, in particular epistemology and the possibility of a metaphysical and mystical understanding of knowledge.

What is the significance of LoF to disciplines besides logic? Some editors and contributors to this journal, notably Dirk Baecker, Stafford Beer, Heinz von Foerster, Louis Kauffman, Niklas Luhmann, and Francisco Varela, have written about the significance of LoF to such diverse subjects as social systems, systems theory and communication, topology, self-reference, natural numbers, electronic circuits, imaginary values in logic, automata, biology, autopoiesis, three-valued logic, and cognitive systems. I refer the reader to them directly regarding the significance of LoF to those disciplines. For my own part, in my short 1996 article "Laws of Form and the Mystical Void", re-published here as an appendix,

I address the significance of LoF to science in general, proposing that LoF might help re-align science with wholeness and spiritual ideas.

Regarding the significance of LoF to epistemology—the nature and origin of knowledge—one of Spencer-Brown's motives in LoF was

> ...bringing together the investigations of the inner structure of our knowledge of the universe, as expressed in the mathematical sciences, and the investigations of its outer structure, as expressed in the physical sciences. Here the work of Einstein, Schrödinger, and others seems to have led to the realization of an ultimate boundary of physical knowledge in the form of the media through which we perceive it. It becomes apparent that if certain facts about our common experience of perception, or what we might call the inside world, can be revealed by an extended study of what we call, in contrast, the outer world, then an equally extended study of this inside world will reveal, in turn, the facts first met with in the world outside: for what we approach, in either case, from one side or the other, is the common boundary between them. ...
>
>   [In the chapter 11], I break off the account [of developing mathematical archetypes] at the point where, as we enter the third dimension of representation with equations of degree higher than unity, the connexion with the basic ideas of the physical world begins to come more strongly into view. [LoF 1972, pxxi-xxii]

Spencer-Brown is proposing a common boundary between the inner mathematical structure of our perception and the outer world of physical science, a boundary which is the archetypal form of our knowing. I use the word "archetype" in the square brackets in the above quotation in order to incorporate an earlier passage where he

> suggests that we have a direct awareness of mathematical form as an archetypal structure. [LoF 1972, pxx]

In his notes to chapter 11, Spencer-Brown continues to describe his vision of how "basic ideas of the physical world" emerge out of the mathematical archetypes outlined in LoF:

> ...we do not imagine the wave train emitted by an excited finite echelon to be exactly like the wave train emitted from an excited physical particle. For one thing the wave form from an echelon is square, and for another it is emitted without energy. (We should need, I guess, to make at least one more departure from the form before arriving at a conception of energy on these lines.) What we see in the forms of expression at this stage...might be considered as simplified precursors of what we take, in physical science, to be the real thing. [LoF, p100-101]

Thus Spencer-Brown considers that certain expressions in chapter 11 of LoF are non-material, energy-less precursors of material particles, precursors in the sense that extending the LoF system further would arrive at conceptions of energy, mass, other physical entities. For a brief discussion of "departure[s] from the form" and of "precursors", see [Engstrom 1999, pages 33-35].

What is the significance of LoF to the possibility of a metaphysical and mystical understanding of knowledge? Much of the significance has to do with the facts [Engstrom 1999, pp 36-40] that the primary ground of LoF is unmarked space, and that forms can not only be created (constructed) out of this unmarked space, but may also be voided (deconstructed) back into this unmarked space:

> LoF offers a unique contribution both to current mathematics and other disciplines which might model relationships involving wholeness and parts. That contribution is a vision in which forms and relations arise from wholeness and can disappear back into that wholeness

("voidability of relations" LoF, p104). To be whole means to be undivided, complete, hale. Mathematics and other notations deal with relations: generally relations between different parts. In the deepest analysis, parts cannot be taken as given; parts must ultimately be generated by dividing wholeness. [Engstrom, 1996a, App. 2 of this paper].

Although this wholes-and-parts relationship may be in two-dimensional mental space, or physical space, or conceptual space, it can also be in metaphysical or mystical space, and in these deeper interpretations unmarked space can represent metaphysical or mystical wholeness, the source of all creation. (See [Engstrom, 1996a, Appendix 3 of this paper], also [Engstrom 1999, pp 33-35].)

In conclusion, the vision of LoF starts from a metaphysical or mystical wholeness, unfolds into forms for logic, and thence toward forms in the physical world. LoF notation embodies this vision to some extent. [Engstrom 1999, pp 33-45]

3. Logic and notation primer

In this section we take leave of the above philosophical tack and prepare for the more technical part of the paper.

Why a logic primer?
Both Peirce's existential graphs and Spencer-Brown's LoF are notations for logic. Since the reader is not expected to be familiar with symbolic logic, in this section we introduce logical and notational concepts, technical terms, and symbols that will be used later on in the paper.

What is logic?
Logic is a subject matter, as are arithmetic and music. Like arithmetic and music, the subject matter of logic deals only with certain kinds of entities, and these entities may have certain values, and these values relate to each other with certain regularities or laws. The entities of arithmetic are numbers. Numbers may have values such as two or four. And laws hold, such as "two times two equals four". This law says that the expressions "two times two" and "four" have the same value, namely four. The entities of music are tones, there is the law of the octave, etc. The basic entities of logic are propositions, or statements. Propositions may be true or false...that is, true and false may be seen as the *values* a proposition may have. For example the statement "two times two equals three" is false, or has the value "false". And the statement "monkeys eat peanuts" presumably has the value "true". But whereas there are many numerical values (1, 2, 3, etc.), in logic there are only two: true and false...these are called "truth-values". Now in logic, more fundamental than the relation of equality is that of *inference* or *implication*, which is sometimes stated in an "if...then..." form: "*If* a particular thing is true, *then* something else is also true." In this case the "particular thing" *implies* the "something else". And *from* the "particular thing" one *can infer* the "something else". In logic, as in arithmetic, certain relations are valid or hold true. One such valid logical implication or inference is that "If all men are animals and if all animals are mortal, then all men are mortal".

Notation

Like arithmetic and music, the subject matter of logic may be conveyed by or expressed in different notations. In arithmetic there are arabic numerals such as 4 and roman numerals such as IV. They have the same meaning or interpretation, namely the number four. By *notation* I mean a system of expressions, together with the rules of their transformation (laws). Thus in the usual notation for arithmetic, 4 and 2x2 are expressions, and 2x2=4 is a law.

In this paper we encounter two different generic kinds of notations. These two kinds of notations are used both for logic and also for relations between classes, about which we will say more later.

• *typographical*[2] notations such as the line of text you are now reading, or as in "((x)⊃y)⊃z", where each occurrence of a letter or other symbol occupies a unique position in a *linearly ordered* sequence read from left to right. The logical notation most commonly used, both in Peirce's day and in our own, is of the "((x)⊃y)⊃z" sort.

• *graphic* notations which "spread out into two dimensions", such as graphs and diagrams, as in [x|y|z]. Graphic notations, with their two-dimensionality, allow more complicated notational connections than do typographical notations. On the other hand there may be no unique sequence, so that the linear ordering taken for granted in typographical notations may be lost. For example, in the two dimensional arrangement ⋰⋰ there is no unique linear path through the dots, no obviously-intended linear ordering of the symbols. Peirce's existential graphs, LoF expressions, and Venn diagrams are all of this graphic kind, and indeed [x|y|z] is a valid expression in all three of those systems.

Logical calculations

As in the algebraic notation for arithmetic, where a letter such as "x" or " A " stands for a numerical value, in logic a letter may be used to stand for *the truth-value* (namely "false" or "true") *of* a some specific or generic proposition. For example we may let the letter "m" stand for the specific proposition "monkeys eat peanuts". Presumably monkeys do eat peanuts, in which case "m" has the value "true". And as with numbers, a letter may stand for a generic value, as in the case where the value is unknown. (Here, the unknown value must be true or false.)

Now just as in arithmetic where "=" means "equals" and "+" means

---

[2] Note: In the word "typographic", the suffix "graphic" refers to "written", not to "pictoral".

addition, logical relations and operations are also represented by special symbols such as $\wedge$, $\vee$, and $\supset$, to use modern symbols. But what are these logical relations and operations?

- At the heart of logic is a negation operation "not" which acts as follows: "not true" is the same as false, and "not false" is the same as true. Combining these gives that false=not(not true) and true=not(not false). And given any proposition p, not(not p)=p.

Given *two* propositions p and q, their truth-values may combine in various ways to make another proposition, just as two numbers may add, subtract, or multiply to make another number. The resulting proposition itself then has a truth-value of either true or false.

- One kind of combination of two propositions is called their *conjunction*, or " p and q", or p$\wedge$q. Its value is false only when at least one of its components is false; otherwise it has the value "true".

- Another kind of combination of two propositions is called their *disjunction* or *alternation*, or "p or q", or p$\vee$q. Its value is true only when at least one of its components is true; otherwise it has the value "false".

- Still another kind of combination of two propositions is *the hypothesis that* p implies q. It is symbolized p$\supset$q, and is called a *conditional* proposition. One may see two reasons for this name. First, because "p implies q" may be rewritten "*if* p, then q", and the q may be seen to be conditional or dependent upon the p ("*if* p..."). Second, viewed as a conditional, the statement "p implies q" is not taken as an assertion of fact, but as a hypothetical or conditional proposition, whose truth or falsity must be evaluated (calculated) from that of p and q. Because, as we mentioned earlier, "p implies q" asserts that "*If* p is true, *then* q is also true", it only has value false (it is only false) when p is true *and* q is *not* true, that is, q is false. More terminology: in a conditional p$\supset$q, the first proposition, namely the first term, here "p" is called *antecedent* or *premise*. The second proposition, namely the second term, here "q" is called the *consequent* or *conclusion*.

There are two kinds of statements that are kin to but subtly different from the conditional proposition. These are "tautological-implication" and "universal propositions". First, tautological-implication:

- A particular conditional proposition p$\supset$q may be true or false, depending upon whether the individual terms p and q are true or false, according to the rule given above. But sometimes, for particular p and q, p$\supset$q has the value "true" under all possible combinations of truth-values for p and q. In this case, p$\supset$q is called a *tautology*, and we say that "p *tautologically-implies* q". This is the technical term for what we mean when we assert as a fact that "p logically implies q" or more simply "p implies q" in a logical context.

A very simple example of a conditional that is a tautology is "p$\supset$p" (namely p$\supset$q where q=p). To see that p tautologically-implies p, we recall that the only possible values for p (that is, for *any* proposition) are true and false, so that there are exactly two possible cases of p$\supset$p, namely "false$\supset$false" and "true$\supset$true". Recalling the rule given above for calculating the value of a conditional, and applying the rule to both cases, we see that "false⊃false" and "true⊃true" both have the value true, since p⊃q only is false when the antecedent is true *and* the consequent is false, neither of which is the case. Since these cases are not false, they are true, as mentioned earlier. Thus, p⊃p has the value true under all possible values of antecedent and consequent, so p⊃p is a tautology.

Looking inside of propositions to find individuals or classes
Now for universal propositions. The statements "all men are animals" and its equivalent "every man is an animal" are examples of *universal propositions*. In order to see the structure of universal propositions, we need to "look inside of" a proposition, something we have not done so far.

But what is a proposition? A proposition is a sentence that is either true or false, along with the actual value of truth or falsity of that sentence. So a proposition is a sentence. And such a sentence may have a subject and an object. Now let us briefly consider yet another subject matter, namely that of *classes* and their individual members. For example, an individual man is a member of the class "men". In our notation for logic thus far, single letters such as "x" or "A", have referred only to propositions or their truth-values. Let us now also use a single letter such as "x" or "M" to refer to either an individual, such as " a man", or else a *class* of individuals, such as "men". Then universal propositions are in general of the form: "all A are B", or "every A is a B".

On the one hand, a universal proposition can be seen as a disguised conditional by rewriting it as "if something is an A, then that something is a B". The conditional is "if a, then b", where the antecedent "a" is the proposition "something is an A" and the consequent "b" is the proposition "something is a B" and the two "somethings" are the same. The "A" and "B" are both classes, and the "something" is an individual.[3] We note for later reference that in terms of classes, the statement "something is an A" means "some individual 'x' is a member of class A", and that this relation is symbolized "x∈A", where "∈" means "is a member of".

On the other hand, universal propositions can also seen to be subtly different from conditionals in their use of the words "all" and "every", which in modern terms is said to *"quantify"* the conditional: *"For every* something, if that something is an A, then that something is a B". Such a "quantified" logic is a more sophisticated, both conceptually and notationally, than the simpler logic of "unquantified" logic of propositions we have considered thus far.

"Barbara", class-inclusion, transitivity, duality
A tautological-implication may involve universal propositions, as in the valid

---

[3] Note that the values true and false do not apply to a class (A or B) nor an individual (x), because they are nouns rather than sentences. By contrast, the antecedent "a", the consequent "b", and the conditional "a⊃b" are all propositions—sentences—and are each either true or false.

logical implication mentioned earlier, that "If all men are animals and if all animals are mortal, then all men are mortal". The general form of this, "If all A are B and if all B are C, then all A are C", is traditionally called "*Barbara*", for reasons that will not concern us here. "Barbara" has the form "If p and if q, then r", where all three terms, p, q, and r, are universal propositions, and may be put in the form "s tautologically-implies q" by letting s be p∧q, namely the conjunction "p and q".

Since we have alluded to the subject matter of *classes*, we point out that the logical relation "p tautologically-implies q" of propositions p and q has an intimate connection with the following equivalent relations of classes P and Q: " P is included in Q", or "P is a subclass of Q", both symbolized as "P⊆Q". The relation of classes "P is included in Q" is analogous to the relation between numbers "x is less than or equal to y", symbolized as "x≤y". This numerical relation "is less than or equal to" is true if either "x is less than y" or if "x is equal to y" so that both "1≤2" and "2≤2" hold true. The symbols ⊆ and ≤ even look alike. The converse of "P is included in Q" and "P⊆Q" is "P includes Q", symbolized as "P⊇Q".

Some relations have a property called *transitivity*. An example of a transitive relation is that of equality "x=y", as in "(2+2) = 4", where x is 2+2 and y is 4. Equality is transitive (has the property transitivity) because: knowing that 2+2=1+1+1+1 and that 1+1+1+1=4 we also know that 2+2=4. We can see the more general structure by using (arbitrary) letters as the terms: "x=y and y=z tautologically-implies x=z". If we now replace the equality symbol "=" by a symbol "R" to stand for *any* given relation, this gives "xRy and yRz tautologically-implies xRz". Any relation having this property is said to be transitive. It turns out that the logical relations of conditional and tautological-implication are transitive, as is the class-inclusion relation "⊆".

We also mention, for later reference, that a rather technical relation exists in the structure of logic called "*duality*", which will surface at the end of the section on LoF, and in the appendix on a LoF-like modification of Peirce's notation.

4. *Laws of Form* primer

( ) , or unmarked space

Traditionally, the values in logic are "true" and "false", but Spencer-Brown's quest led him deeper and deeper until elegance forced him to do away altogether with any symbol for the second value—one value he marks with a symbol, the other he leaves unmarked. He discovered the value of *unmarked* space and its power for a notation, analogous to the concept of zero, but more subtle and profound in the sense that it is more pervasive and primordial (see the section on the significance of LoF). Most notations start with a blank sheet of paper, or an equivalent such as a computer screen showing no symbols, but these blank spaces have no interpretation until symbols are placed in them and they no longer blank. But in LoF unmarked space—whether a blank sheet or any part (area) of the sheet without a mark—is both a LoF expression and also one of the two LoF values. One may intuitively appreciate this use of unmarked space as a value and expression by realizing that "doing nothing" is a course of action and hence "is doing something".

Boxes, unmarked, and very little else.
Not only does LoF eliminate any symbol for the second value, but Spencer-Brown also found that for logical calculations he could dispense with *all* of the symbols (such as "false", ∨, ∧, ⊃, and parentheses) except variables and one for negation, if negation were represented by a box, or circle or other such topologically simple closed curve. (Peirce also used boxes, circles, and closed curves, as well as unmarked space, as we will see in later sections.) The box "☐" has a boundary which creates and encloses a space or area inside it that is distinct or severed from the space outside. Whereas in most mathematical and logical notations parentheses are necessary to construct many expressions, in LoF parentheses or brackets are never needed, partly due to the fact that the boundary of each LoF cross already serves to bracket its inside area.

All expressions in LoF notation are built up from just unmarked, boxes, and variables *via* side-by-side juxtaposition and/or enclosure-within-boxes. Expressions (variables, boxes, unmarked space, or more complicated combinations of these) can be placed side-by-side, just as in standard notations: for example "xy", or "x☐", or "☐☐". In the last example just shown, each box is outside the other. But since each box is also an enclosure, it has room for symbols to stand in the space inside it or be "nested" or enclosed inside it: " x̄ ". Given any LoF expression x, we will call the expression x̄ call the *negation* of x. In the example x̄, if x = ☐ then we have two nested boxes "☐̄". The size of a box is irrelevant, and boxes are not allowed to touch or intersect each other. So a typical LoF expression might look like this: $\overline{\overline{x} y \overline{} z}$.

In LoF notation expressions in the same space have no sequential order. For example, here are five arrangements of a and b; all are equivalent: $a\,b = {}_a b = \frac{b}{a} = {}^b a = ba$. This contrasts with typographical notation, in which each symbol occupies a unique position in a line read from left to right...just as in the text you are now reading. In typographical notation, "ab" and "ba" are *different* expressions because the sequential order of the "a" and "b" are different, and in typographical notation "ab" and "ba" are generally *not equivalent*. But in LoF expressions there is no sequence in terms of side-by-side position. The expression

⌐a⌐ ⌐b⌐ c, for example, is identical to ⌐a⌐ c ⌐b⌐, to ⌐b⌐ ⌐a⌐ c, and to $\boxed{\dfrac{\boxed{a}}{\boxed{b}}\;c}$ because all of these arrangements are characterized by the following "inside/outside" relationships: c stands in the common space outside the two boxes while a and b each stand inside a different box. But there *is* a restriction to rearrangements in LoF notation: namely that neither boxes or any other expressions are allowed to touch or intersect each other; consequently, "no box-boundaries may be crossed". For example, the expression ⌐a⌐ ⌐b⌐ c is *not* equivalent to ⌐ ⌐ a ⌐b⌐ c, nor to ⌐ ⌐ab⌐ c,[4] because the "inside/outside" relationships are all different. To rearrange ⌐a⌐ ⌐b⌐ c into the other two, the a in ⌐a⌐ would have to cross a boundary of a box, and this is forbidden.

The only ordering, a partial one, in LoF expressions is by what is called *depth of space*. The space inside a box is said to be one space *deeper* than the space outside it. This ordering by depth of space is analogous to the use of parentheses in typographical notation, which, like LoF boxes, may be nested.

This completes our primer of LoF's expressions, except to note that Spencer-Brown abbreviates the boxes ⌐ ⌐ by showing only the top and right sides thus: ⌐; which he calls a "cross". For example, the LoF expression ⌐x⌐ y ⌐ ⌐ z is abbreviated as ⌐x⌐ y ⌐ z.

LoF postulates and calculations

For algebraic calculations, LoF uses two postulates and nine derived consequences numbered C1 through C9. Three of these will concern us here: C1, C2, and C5: C1 or "reflexion" is $\overline{\overline{a}}$ = a; C2 or "generation" is $\overline{ab}$ b = $\overline{a}$ b; and C5 or "iteration" is a a = a. Recalling that one possible value for a variable is: "unmarked", this special case of C1, $\overline{\overline{}}$ = unmarked, is called I2 or "order".

Three LoF interpretations as logic

As for the meaning or interpretation of LoF notation, Appendix 2 of LoF shows three different interpretations as logic, which I will simply call "interpretation-A", "interpretation-B", and "interpretation-C". In all three interpretations each

---

[4] Note: the letters a and b in ⌐ab⌐ are not touching, as there is some unmarked space separating them.

variable denotes some sentence which must be either true or false, and the negation of any expression c is represented by its enclosure $\overline{c|}$ in a cross, that is, in a box.

Interpretation-A and interpretation-B are opposites in the sense that in the former, unmarked denotes the value "false", while in the latter unmarked denotes the value "true". Directly below we list how to interpret some simple expressions.

Logic	LoF Interpretation-A	LoF Interpretation-B
false	unmarked	$\overline{|}$
true	$\overline{|}$	unmarked
not-p	$\overline{p|}$	$\overline{p|}$
p and q	$\overline{\overline{p|}\;\overline{q|}|}$	p q
p or q	p q	$\overline{\overline{p|}\;\overline{q|}|}$
if p, then q	$\overline{p|}$ q	$\overline{\overline{q|}\;p|}$

(It turns out that interpretations-A and -B each bear the technical relation of "dual" of the other.)

In interpretation-C, interpretation-A is extended to apply to inferences such as "If all men are animals and if all animals are mortal, then all men are mortal" ("Barbara"). In this case a letter such as "p" stands for the proposition " a particular individual, x, is a member of class p". The fact that the logical relations depend upon the word "all" makes them "quantified", and beyond the scope of interpretations-A or -B, which are "unquantified".

5. Features of *Laws of Form* notation anticipated by Peirce

With the preliminaries done, we now come to the main part of the paper: Peirce's precursors to LoF. That task will occupy this and the following three sections. Here in this section we identify certain LoF notational features. In the next three sections each their own subsections, we show Peirce's precursors to those features. In the subsections, after a brief preview of existential graph terms and symbols we let Peirce describe his own notation, with quotations—sometimes long quotations—from Peirce's *Collected Papers*. In the process, relevant parts of Peirce's existential graphs and other work are explicated.

We begin by listing below some features of LoF that the alpha part of Peirce's existential graphs, and in some cases also earlier work by Peirce in other notations, foreshadow. This list is not intended to be exhaustive. We will label the features so that we can refer to them as we proceed. Each feature is denoted by a letter that somehow relates to its meaning: such as "u" for unmarked, "d" for dimension or depth, etc.

Features of expressions:
(u.1) Unmarked space as an expression;
(b.1) The LoF cross (abbreviating a "box" or rectangle or any closed curve) as a boundary which may enclose other symbols;
(b.2) The LoF cross (abbreviating a "box" or rectangle or any closed curve) as an area/boundary structure;
(d.1) Unordered two-dimensional notational space;
(d.2) Ordering by "depth" of two-dimensional notational space, not by side-by-side position within the notational space;
(w) The above features together—expressions consisting *only* of unmarked space, enclosure-boundaries, and variables such as A, B, C, or x, y, z—form an integral semiotic whole. (Semiotic has to do with signs, symbols...what we here call expressions.)

Features of interpretation:
(t) Interpreting unmarked space as the value "true";
(j) Interpreting juxtaposition as the logical operation of conjuction: ∧;
(n) Interpreting enclosure as negation;
(np) Condensing or conflating signs for negation and parentheses, ultimately via a graphic icon of enclosure such as a rectangle ("box") or oval;
(c.1) Conditional and universal propositions can be expressed in the same form, namely $\overline{p}$ ∨ q or p⊃q;
(c.2) Interpreting "$\boxed{A\boxed{C}}$" as a conditional or universal proposition.

Features of the laws:
As for the laws, here is a list of some algebraic laws of LoF that Peirce's "Illative [or Inferential] Transformations" of existential graphs foreshadow:
C5 iteration;
C2 generation;
C1 reflexion and I2 order;
where the letter-with-number designations (C5, etc.) as well as the names (iteration, etc.) are Spencer-Brown's.

Miscellaneous feature:
(u.2) Unmarked space as pervasive space.

6. LoF ideas in Peirce's pre-existential-graph work

Now we start finding passages in Peirce's *Collected Papers* that foreshadow LoF's features. Existential graphs are only a fraction of the work in logic by Peirce, so we will find some foreshadowings in his earlier work: in his symbolic logic in typographical notation, and in his first attempt at logical graphs: his "entitative graphs".

Feature (np): condensing negation and parentheses

One of Peirce's earliest steps in the direction of LoF, was feature (np), that of conflating or condensing signs for negation and parentheses into a single sign. He does this successfully for the conditional proposition x⊃y, where the first term, x, is called the antecedent and the second term, y, is called the consequent, as shown in the quotation below.

[3.389-Published 1885][5] We have seen that x⊃(y⊃z) may be conveniently written x⊃y⊃z; while (x⊃y)⊃z ought to retain the parenthesis. Let us extend this rule, so as to be more general, and hold it necessary *always* to include the antecedent in parenthesis. Thus, let us write (x)⊃y instead of x⊃y. If now, we merely change the external appearance of two signs; namely, if we *use the vinculum instead of the parenthesis*, and the sign v instead of ⊃, we shall have

x⊃y	[i.e. (x)⊃y]	written $\overline{x}$ v y
x⊃y⊃z	[i.e. (x)⊃(y)⊃z]	written $\overline{x}$ v $\overline{y}$ v z
(x⊃y)⊃z	[i.e. ((x)⊃y)⊃z]	written $\overline{\overline{x}\ \text{v}\ y}$ v z, etc.

... and the vinculum becomes identified with the sign of negation.[6]

The key steps in the quotation are that 1. he says "*always* to include the antecedent in parenthesis", hence "let us write (x)⊃y instead of x⊃y"; 2. "we *use the vinculum* [to represent] *the parenthesis*"; 3. "*the vinculum becomes identified with the sign of negation*" ...so that the vinculum now, per 2 and 3, serves as the sign of *both* parenthesis and of negation simultaneously, as in the example $\overline{x}$ v y v z. There are two details to note to make sense of the [3.389] passage quoted. First, Peirce is already using the vinculum specifically for negation earlier in the same chapter: for example, where he says "1–x [which equals negative x] is best written $\overline{x}$ " [3.370; see also the second footnote]. Second, Peirce's replacement of ⊃ by v in the context of step 3 ("use the vinculum instead of the parenthesis, *and the sign v instead of ⊃*"—my italics) is valid because the conditional "x⊃y" can be re-written in terms of negation and disjunction as "(not-x)-or-y", or $\overline{x}$ v y.

So, in [3.389] the vinculum serves as the sign of *both* parenthesis and of negation simultaneously, as in the example $\overline{x}$ v y v z. To see how close this is to LoF notation, we have only to see each vinculum as the top of a box, and to extend it into the entire box thus: $\boxed{x\ \text{v}\ y}$ vz and delete the " v " operation sign, thus:

---

[5] Recall our convention that if no author or title is indicated in square brackets, then it refers to Peirce's *Collected Papers*, where for example, [3.389] means Volume 3, paragraph number 389.

[6] Note: Here and throughout the paper I have replaced his signs for the conditional operation and disjunction by the modern symbols ⊃ and v, respectively.

$\boxed{x|y}$ z, which in interpretation-A of LoF is the proper transcription of $\overline{x}$ ∨y ∨z.

Indeed this successful conflation by Peirce of signs for negation and parentheses continues intact in Peirce's existential graphs. For example, in [4.378-published 1911], Peirce uses typographical bracketing (square brackets, parentheses, and braces) as abbreviations for enclosing a term in an oval line or [Fig. 65, 4.383-published 1911] rectangle, and asserts that any such enclosure has the same meaning of negating what it encloses: "the square brackets and parentheses precisely deny (or negate) what they enclose...Braces may also be used for the same purposes".

Feature (c.1): conditional = universal proposition
In [3.444-3.445-Published 1896], Peirce explicitly asserts feature (c.1), that a conditional proposition x⊃y has the same logical structure or form, namely $\overline{A}$ ∨B, as a universal proposition "every A is B."[7]

> [3.444-3.445]: [The conditional proposition, A⊃B, is] 'if A is True, then B is true'... or, what is the same thing, 'Either A is not true, or B is true' [$\overline{A}$ ∨B]... Now let us express the categorical [i.e., universal] proposition, 'Every man is wise'... that is... either m_i is not true or w_i is true [where] m_i is the sentence 'any individual object *i* is a man' and w_i is the sentence 'any individual object *i* is wise' [$\overline{m_i}$ ∨w_i]... The conditional and categorical [i.e. universal] propositions are expressed in precisely the same form...the *form* of relationship is the same [i.e., $\overline{x}$ ∨y, or x⊃y].

Similarly, in interpretation-C of LoF the universal proposition "*all a are b*" is cast into the form of the conditional proposition a⊃b, but instead of writing, for example, the antecedent *a* as "m_i" as Peirce does, Spencer-Brown writes x∈a (this would write be i∈m in terms of Peirce's letters) and then abbreviates x∈a to *a* (and similarly with the consequent *b*):

> [LoF p119]: All universal forms of the traditional logic of classes can be accommodated within the logic of sentences...To accommodate them we use the pattern in the following key.
>
> for *all a are b* use (x∈a)⊃(x∈b)
>
> ...To avoid the use of distinct letters for sentences and classes, we can allow, in the calculating forms, any simple literal variable *v* to stand for the sentence x∈v, i.e. 'x is a member of the class *v*'. This will not lead to unintentional confusion, since the sign *v*, as used to denote the class, does not enter into the calculation, which is undertaken with *v* representing only the truth value of the corresponding sentence.

Thus, (x∈a)⊃(x∈b) becomes a⊃b; that is, the universal proposition *all a are b* is expressed in precisely the same form as the conditional proposition, namely a⊃b.[8]

---

[7] Peirce's assertion is about logical structure independent of notation, and it holds in his existential graphs as well.

[8] Note: In modern notation a universal proposition normally explicitly requires the quantifier "for all x...", yet both Spencer-Brown and Peirce are here getting by without such explicit quantification.

Feature (b.1): a closed curve as a boundary which may enclose other symbols; and 'feature (n)': interpreting enclosure as negation
So far, all of Peirce's work cited regarding the above two features utilize typographical notation rather than his "graphs". But even earlier than his existential graphs, which will be our main focus, were his entitative graphs: published 1897.

In his entitative graphs, Peirce uses a closed curve to enclose other symbols (feature (b.1)). In [3.475-Published 1897], Peirce represents the universal statement 'all men are mortal' by the "logical graph" (entitative graph) $\boxed{h}\!-\!d$, where "the line that joins antecedent [h] to consequent [d] encircle[s] the whole of the former." (The enclosing loop is rendered here as a box, for the author's convenience.) And, tellingly, he says "...the antecedent...has a negative...character. So already Peirce uses an enclosure to denote negation (feature (n)), but only secondarily: it's primary function being to identify the antecedent.

But later in the same paper [3.493-Published 1897], he *reinterprets* the enclosure such that negation is now its full meaning:

> I entirely discard the idea [of antecedent and consequent]...I consider the circle [enclosure] round the antecedent as a mere sign of negation, for which in the algebra I substitute an obelus [he uses "obelus" and "vinculum" interchangebly here] over that. The line between antecedent and consequent I treat as an operation [v, in modern symbols]. [3.493-Published 1897]

That is, $\boxed{h}\!-\!d$ = $\overline{h}$ v d. That negation is now enclosure's full meaning is indicated by the word "mere" in the quotation.

So here we have in his entitative graphs an enclosure as the negation of what it encloses, thus combing features (b.1) and (n).[9],[10] Nor does Peirce seem to think of the circle as creating a separate and independent area inside, as evidenced by his word "mere" in the statement, above, that "I consider the circle...as a mere sign of negation, for which in the algebra I substitute an obelus over that."

---

[9] His statement "I entirely discard the idea [of antecedent and consequent]" can be understood as meaning that he dissolves the *asymmetrical* distinction between antecedent and consequent in the conditional form (of implication) "h⊃d" into its equivalent *symmetrical* disjunctive form of "(not-h)-or-d", or $\overline{h}$ v d. Technically the symmetry is called commutativity, but the point is that the sequential order does not matter: that is, $\overline{h}$ v d equals d v $\overline{h}$. This symmetry is part of Feature (d.1): two-dimensional notational space is not sequential, to be discussed later. By contrast, the order *does* matter in h⊃d: "d⊃h" is not equivalent to it.

[10] Because, in $\boxed{h}\!-\!d$, he retains the horizontal straight line or dash for the relation with d, he has not given up the "binary scope" of his relations. The "scope" of a relation is what it applies to. "Binary" scope means that it applies to two terms. For example, the plus sign "+" applies to two numbers, as in "2+3". In some later notations, Peirce gives up the binary scope of his relations ...see feature (d.1) in [4.374-published 1911]), and LoF specifically does this: e.g. LoF p89, p92, p109).

In a later paper, [4.378-published 1911], Peirce integrates his notation of graphs with typographical notation by using typographical bracketing (square brackets, parentheses, and braces) as abbreviations for enclosing a term in an oval line or [Fig. 65, 4.383-published 1911] rectangle, and asserts that any such enclosure has the same meaning of negating what it encloses: "the square brackets and parentheses precisely deny (or negate) what they enclose...Braces may also be used for the same purposes".[11]

In light of Peirce's eventual [1911] equivalence, described in the paragraph directly above, of parentheses and enclosure, both as negation, we may see a retrospective significance to his use of parentheses in the early work [3.389-Published 1885] already cited: "Let us hold it necessary *always* to include the antecedent in parenthesis. Thus, let us write (x)⊃y instead of x⊃y ... [he later rewrites this as $\overline{x}$ ∨y]." We can see in this, his 1885 uses of parentheses in (x)⊃y and of obelus-or-vinculum in $\overline{x}$ ∨y (i.e. "we *use the vinculum* [to represent] *the parenthesis*" [3.389]) *the same impulse* as in the 1897 use of a circle in $\boxed{h}$–d and of circles, ovals, and rectangles in his later existential graphs: that of placing the antecedent in its own separate space, whether typographic as in the cases of parentheses and of vinculum, or graphic in the case of circles or other enclosures, all of which have the meaning of negation. This is feature (n), that of interpreting enclosure in a separate space as the form of negation.

7. Features of *Laws of Form's* expressions and interpretation in Peirce's existential graphs

In this, the largest, section we finally come to to Peirce's *existential* graphs themselves ([4.618-published 1908]: "invented in January 1897 and not published until October, 1906"). In the first subsection we begin with a preview comparing terms and expressions in LoF with those of "the alpha part" of existential graphs. For discussion of the "alpha" and other parts of existential graphs, see subsection "Feature (w)..." in this section. In the remaining subsections we cite and discuss passages in the *Collected Papers* that foreshadow a number of features of LoF.

Introducing Peirce's existential graph notation

Since we already have some acquaintence with LoF notation, we introduce Peirce's existential graphs by comparing existential graph expressions and terms with similar LoF expressions and terms, in the following table.

LoF	Existential graphs
expression	graph
unmarked space	blank, sheet of assertion,

---

[11] We also note that in [4.378-published 1911] he also accomplishes "feature (np)", of successfully conflating signs for negation and parentheses.

"cross" or ⏋ as boundary of box ☐	☐ as cut, loop, oval, sep, enclosure-boundary,
⏋ as box ☐ including inside	☐ as enclosure
variable expression, such as p	graph p
p⏋ as ☐p☐	☐p☐
reflexion ☐☐ or ☐p☐	double cut ☐☐ or ☐p☐
q⏋p as ☐q⏋p☐ or ☐p⏋q☐	☐p☐q☐
p q	p q as conjuction

An example of ☐p☐ is ☐X is good☐ [4.461-Unpublished c.1903]. An example of ☐p☐q☐ is ☐Antecedent☐Consequent☐ [4.436], or more concretely ☐It hails☐It is cold☐ [4.528], where again, only the depth of space is important and not the left-right-up-down order.

Features (b.1): a closed curve as a boundary which may enclose other symbols; and (b.2): a closed curve as an area/boundary structure

Continuing now our thread from the entitative graphs, the existential graphs clearly have feature (b.1), a closed curve as a boundary which may enclose other symbols. The boundary of the enclosure, be it a rectangle ("box") or loop or circle or oval, is called a "cut" (or sometimes "sep", as in "separate", from the Latin word "sæpes" or "saepes" for fence [4.435]). While retaining its interpretation as negation (see the subsection on feature (n) below), which it inherited from his earlier entitative graphs (see the subsection on feature (n) above), in the existential graphs the cut acquires an area/boundary relationship with many of the characteristic LoF properties and much the same "flavor": the LoF box ☐ has a boundary which creates and encloses a *space* or *area* inside it that is distinct or severed from the space outside; the space inside a box is said to be one space *deeper* than the space outside it. (See the subsection on feature (d.2) below); boxes may be nested, thus: ☐☐; and boxes are *not allowed to touch*[12] or *intersect* each other.

[4.414(5)-published 1903]: A *cut* is a self-returning finely drawn line. A cut is not a graph... A cut drawn upon the sheet of assertion severs the surface it encloses, called the

---

[12] Actually, Peirce does sometimes allow loops or circles or ovals to touch (for example [4.436], [4.440], [4.449-4.453], [4.457], [4.461], [4.528], and [4.564]), but that does not alter the spirit of his usage, which is that each loop or box creates a separate space.

area of the cut, from the sheet of assertion; so the area of a cut is no part of the sheet of assertion. A cut drawn upon the sheet of assertion together with its area and whatever is scribed upon that area constitutes a graph...and is called the *enclosure* of the cut. ...A cut can (if permitted) be drawn upon the area of any cut, and will sever the surface which it encloses from the area of the cut, while the enclosure of such inner cut will be a graph... scribed on the area of the outer cut. ...Two cuts one of which has the enclosure of the other on its area and has nothing else there constitutes a *double cut*.

[4.414(6)-published 1903]: No graph or cut can be placed partly on one area and partly on another.

[4.399-published 1903]: By a *Cut* shall be understood to mean a self-returning linear separation (...[or] line) which severs all that it encloses from the sheet of assertion on which it stands itself, or from any other area on which it stands itself. The whole space within the cut (but not comprising the cut itself) shall be termed the *area* of the cut. Though the area of a cut is no part of the sheet of assertion, yet the cut with its area and all that is on it, conceived as so severed from the sheet, shall, under the name of the *enclosure* of the cut, be considered as on the sheet of assertion or as on any other area as the cut may stand upon. Two cuts cannot intersect one another, but a cut may exist on any area whatever. ...A cut is not a graph; but an *enclosure* is a graph.

Thus: a graph is an expression (and can thus be represented by a variable); a cut is the boundary-only of the box ▢ and is not by itself an expression; a double cut is two nested boxes ▣ ; a cut, as a boundary, severs the area inside the box ▢ from the area outside; an enclosure is an expression of the form B consisting of the box-boundary plus the box-interior or area inside, including any graphs (here B) scribed on the area; and finally, because a cut is part of a graph, the statement "No graph or cut can be placed partly on one area and partly on another" ensures that "Two cuts cannot intersect one another".

For continued discussion of feature (b.2), see the final subsection of this section, "Feature (b.2): a "cut" as an area/boundary structure".

Feature (d.2): ordering by "depth" of two-dimensional notational space
Just as the space inside a LoF box is said to be one space *deeper* than the space outside it, Peirce says [4.578-conference 1906]: "the cut may be imagined to extend down to one or another depth into the paper...", thus satisfying feature (d.2), ordering by "depth" of two-dimensional notational space.

Feature (n): interpreting enclosure as negation ("cut" as negation)
In the existential graphs the "cut" retains its interpretation as negation, as the following quote indicates.[13]

[4.474(3)-Unpublished c.1903]: ...a sep [or fence], or...oval [i.e. a "cut"], when unenclosed is with its contents (the whole being called an *enclosure*) a graph... which precisely denies the proposition which the entire graph within it would, if unenclosed, affirm.

Or, graphically, using a box instead of an oval for the cut, [4.461-Unpublished

---

[13] Note: Peirce's synonyms for "cut" include loop, oval, and "sep", from a Latin word for fence.

c.1903]: "Fig. 113 [' $\boxed{\text{X is good}}$ '] ...[is] the precise denial of Fig. 112 ['X is good']." More simply, '$\boxed{\text{B}}$' is the negation of 'B'.

Feature (u.1): unmarked space as an expression
Peirce's existential graphs have feature (u.1), that unmarked space is an expression, or "graph".[14]
> [4.396-published 1903]: It is agreed that a certain sheet, or blackboard, shall, under the name of *The Sheet of Assertion*, be considered as representing the universe of discourse, and as asserting whatever is taken for granted...to be true of that universe. The sheet of assertion is, therefore, a graph.
> [4.397-published 1903]: ...the sheet itself [is] a graph...if the sheet be blank, this *blank*, whose existence consists in the absence of any scribed graph, is itself a graph.

The passages above show that unmarked space is an expression (i.e., graph) when the entire sheet is unmarked (i.e., blank). The following passage shows that unmarked space is an expression even when it is but a part of the sheet. [4.398-published 1903]: "Every blank part of the sheet is a...graph." Again,
> [4.414(4)-published 1903]: The sheet of assertion is itself a graph...and so is any part of it, being called a *blank*.
> [And 4.414(5)-published 1903]: Any blank part of any area is a graph...

More graphically, Fig.202 in [4.564-published 1906] shows a blank or "void" or unmarked area between two nested cuts, which, using our boxes, would look like this: $\boxed{\boxed{\text{C}}}$, namely $\boxed{\text{A}\boxed{\text{C}}}$ where A = unmarked.

Feature (d.1): unordered two-dimensional notational space, with no sequential order
Peirce recognized that when "commutativity" and "associativity" hold (that is, xy=yx and (xy)z=x(yz), respectively[15]) the symbols x, y, z no longer need be confined to the typographic line with its left-to-right sequential ordering:
> [4.374-published 1911] ...Operations of commutation, like xy ∴ yx, may be dispensed with by not recognizing any order of arrangement as significant. Associative transformations, like (xy)z ∴ x(yz), which is a species of commutation, will be dispensed with in the same way: that is, by recognizing an equiparant as what it is, a symbol of an unordered set.

Here x, y, z are all what he calls "equiparants", and "xy ∴ yx" is another way of expressing the equation "xy=yx". Likewise "(xy)z ∴ x(yz)" is another way of expressing the equation "(xy)z=x(yz)".

Moreover, he embodies this unordered-ness in his notation when he says
> [4.378-published 1911] We... express the fact that 'if A can be true, B can be true' by

---

[14] Note: Peirce's term for expression is "graph".
[15] For example, the addition of numbers is both commutative and associative: witness 1+2 = 2+1 because both equal three, and (1+2)+3 = 1+(2+3) because both equal six. It turns out that in logic, commutativity and associativity hold for conjuction and also for disjunction, but not for implication.

$[A(B)]$ or $[(B)A]$ or $\left[\begin{smallmatrix} A \\ (B) \end{smallmatrix}\right]$, etc.

and then directly adds "The arrangement is without significance", by which he means that the order of arrangement of the components A and (B) within the square brackets is without significance, so that A and (B) are free to be anywhere within the square brackets:

$$[A(B)] = [(B)A] = \left[\begin{smallmatrix} A \\ (B) \end{smallmatrix}\right] = \left[\begin{smallmatrix} A \\ (B) \end{smallmatrix}\right] = \text{etc.}$$

Since, earlier in [4.378], Peirce explained that he uses square brackets and parentheses as abbreviations for enclosing a term in an oval line (or rectangle, per Figure 65 in [4.384]), he explicitly means these to be abbreviations of (respectively):

$$\boxed{A\boxed{B}} = \boxed{\boxed{B}A} = \boxed{\begin{smallmatrix} A \\ \boxed{B} \end{smallmatrix}} = \boxed{\begin{smallmatrix} A \\ \boxed{B} \end{smallmatrix}} = \boxed{\begin{smallmatrix} \boxed{B} \\ A \end{smallmatrix}} = \boxed{\begin{smallmatrix} \boxed{B} \\ A \end{smallmatrix}} = \text{etc.}$$

Later, in the section on feature (c.2) we will see that this is Peirce's existential graph for conditional and universal propositions.

In LoF these same expressions would normally be abbreviated as "crosses":

$$\overline{A\,\overline{B}} = \overline{\overline{B}\,A} = \overline{\begin{smallmatrix} A \\ \overline{B} \end{smallmatrix}} = \overline{\begin{smallmatrix} A \\ \overline{B} \end{smallmatrix}} = \overline{\begin{smallmatrix} \overline{B} \\ A \end{smallmatrix}} = \overline{\begin{smallmatrix} \overline{B} \\ A \end{smallmatrix}} = \text{etc.,}$$

but the LoF crosses stand for box-enclosures, just as do Peirce's brackets and parentheses above.

Feature (w): expressions consist only *of variables, unmarked space, and enclosure-boundaries; also: alpha, beta, and gamma parts of existential graphs*

As mentioned in the Abstract and at the end of the Introduction, it is only the alpha part of his existential graphs which will concern us in this paper. We can see why from Peirce's description of these parts, in

[4.512-published 1903]: The alpha part [of existential graphs] has [only] three distinct kinds of signs, the *graphs*, the *sheet of assertion*, and the *cuts*.

These three kinds of signs correspond to feature (w), that expressions consist *only* of variables (which stand for any expression—Cf [4.395]), unmarked space, and enclosure-boundaries, respectively, with the proviso that in LoF unmarked space refers not only to the entire sheet but also to any unmarked area of it...but this is consistent with Peirce's [4.564-published 1906] Fig. 202 which, using boxes, would look like this: $\boxed{\boxed{C}}$, which shows a blank or "void" or unmarked area between two nested cuts, namely $\boxed{A\boxed{C}}$ where A=unmarked.

Peirce's description continues: "The beta part part adds two quite different signs, *spots*, or *lexeis*, and *ligatures* with *selectives*...[also 'lines of identity'.] ...in the gamma part of the subject all the old kinds of signs take new forms..." Since the beta and gamma parts introduce symbols and marks that are not part of

LoF—namely "*spots,* ...*ligatures* , ...lines of identity, ..."—these extensions are not applicable to this paper, as there is no counterpart in LoF.

Feature (t): interpreting unmarked space as a true proposition
Peirce's existential graphs have feature (t), that of interpreting unmarked space as a true proposition; i.e. as the logical value "true".
> [4.396-published 1903]: It is agreed that a certain sheet, or blackboard, shall, under the name of *The Sheet of Assertion*, be considered as representing the universe of discourse, and as asserting whatever is taken for granted...to be true of that universe.

Since initially there is nothing scribed on the sheet, this *blank* sheet (unmarked space) "assert[s] what... is true...". Similarly,
> [4.617-published 1908]: ...vacant cut ["☐"]...[denotes] absurd [i.e. false], and, since

enclosure is negation (i.e. '$\boxed{B}$' is the negation of 'B'), this means that vacant or unmarked denotes the value "true".

In contrast,
> [4.434-Unpublished c.1903]: in my first invented system of graphs, which call *entitative graphs*, a blank sheet instead of expressing only what was taken for granted [i.e. as true] had to be taken as an absurdity [i.e., as false].

Feature (u.2): unmarked space as pervasive space;
In this subsection we take the liberty of speaking less rigorously than in the rest of the paper. Peirce and Spencer-Brown seem to have similar ideas or mentalities regarding unmarked notational space as a background which contains properties which pervade expressions in their space.

One of LoF's principles [LoF p43] is the:
> Principle of relevance. *If a property is common to every indication, it need not be indicated.* ...[For example, unmarked space] is common to every expression in the calculus of indications and ..., by this principle, has no necessary indicator there.

An example outside of his formal system, in daily life, might be in a diary where entries under a given date need not list the date, since that date is already understood to apply to all the entries. It is, metaphorically, as if the unindicated common property is contained within the unmarked space which is the ground on which or background against which expressions or indications stand, and as if the unmarked space along with its unindicated common property "pervades" any and all expressions which might stand in the space. Indeed, Spencer-Brown calls the space in which an expression stands its "pervasive space" [LoF p7].

Somewhat similar to this is the (initially blank) sheet of assertion of Peirce's existential graphs, which implicitly includes all properties, or true statements, of the system.
> [4.396-published 1903]: It is agreed that a certain sheet, or blackboard, shall, under the name of *The Sheet of Assertion*, be considered as representing the universe of discourse, and as asserting whatever is taken for granted...to be true of that universe.

The sheet is common to all graphs placed on it, and thus, as if by Spencer-Brown's principle of relevance, is, itself, blank. Also the blank has the value "true", and so it is as if all of the structure (true statements, or properties) defining the existential-graph system becomes the unmarked background which "pervades" every graph and, by gestalt principles of figure/ground perception, every graph is seen as a figure — i.e. as what-is-relevant-and-indicated standing *out and against* an, unmarked and not-indicated background which is less relevant.

One final point that applies to both systems is that although some property *need not be* indicated, it *may* be indicated at will. In the case of both LoF and existential graphs unmarked space may be replaced by ▢, which Peirce calls a double cut. In the case of LoF, unmarked space may be replaced more generally by |A|A|| (per [LoF p28] J1: [$\overline{a|a}$ = unmarked) or by a myriad of equivalent expressions. Analogously, in existential graphs,

[4.507-Unpublished c.1903]: Rule 3. ...any graph well-understood to be true may be scribed unenclosed.

Feature (j): interpreting juxtaposition as logical conjunction: ∧
Peirce's existential graphs have feature (j), of interpreting juxtaposed graphs "AB" as logical conjunction "A∧B" of propositions A and B, as indicated in the following quotes.

[4.376-published 1911] If...we write two propositions on the same sheet...both are asserted. [Likewise 4.564-published 1906]: AB, which results from setting both [existential graphs] A and B upon the same sheet, shall assert that both...are real [i.e. true].

That is, asserting "both A and B" asserts "A∧B".

To underscore his meaning of "A∧B", Peirce then adds, by way of contrast:
This was not the case in my first system of Graphs, ...which I now call *Entitative Graphs*..., [in which, he explains in 4.434-Unpublished 1903]: propositions written on the sheet together were not understood to be independently [conjunctively] asserted ["A∧B"] but [instead] to be alternatively [disjunctively] asserted ["A∨B"].

Feature (c.2): interpreting |A|C|| *as a form of the conditional or universal proposition; and: existential graphs use LoF interpretation-B*
Existential graphs use interpretation-B of LoF, for which features (t) and (j) and (n) hold, as well as (c.2): namely, that p⊃q is transcribed as $\overline{q|p}$, or $\overline{p\,\overline{q}}$, i.e. |p|q||.

In the quotation below, Peirce gives several reasons why he finds the diagram |A|C|| (as an existential graph) to be an appropriate form to represent the inference "Antecedent implies Consequent" or "A implies C" or "A⊃C".

[4.435-Unpublished 1903]: [For an expression to] be able to express that a...consequent, C, necessarily follows from an...antecedent, A ...the antecedent and consequent must be in separate compartments... In order to make the representation of the relation between them iconic, we must ask ourselves what spatial relation is analogous to their relation. Now if it be true that 'If *a* is true, *b* is true' and 'If *b* is true, *c* is true,' then [it follows logically that]

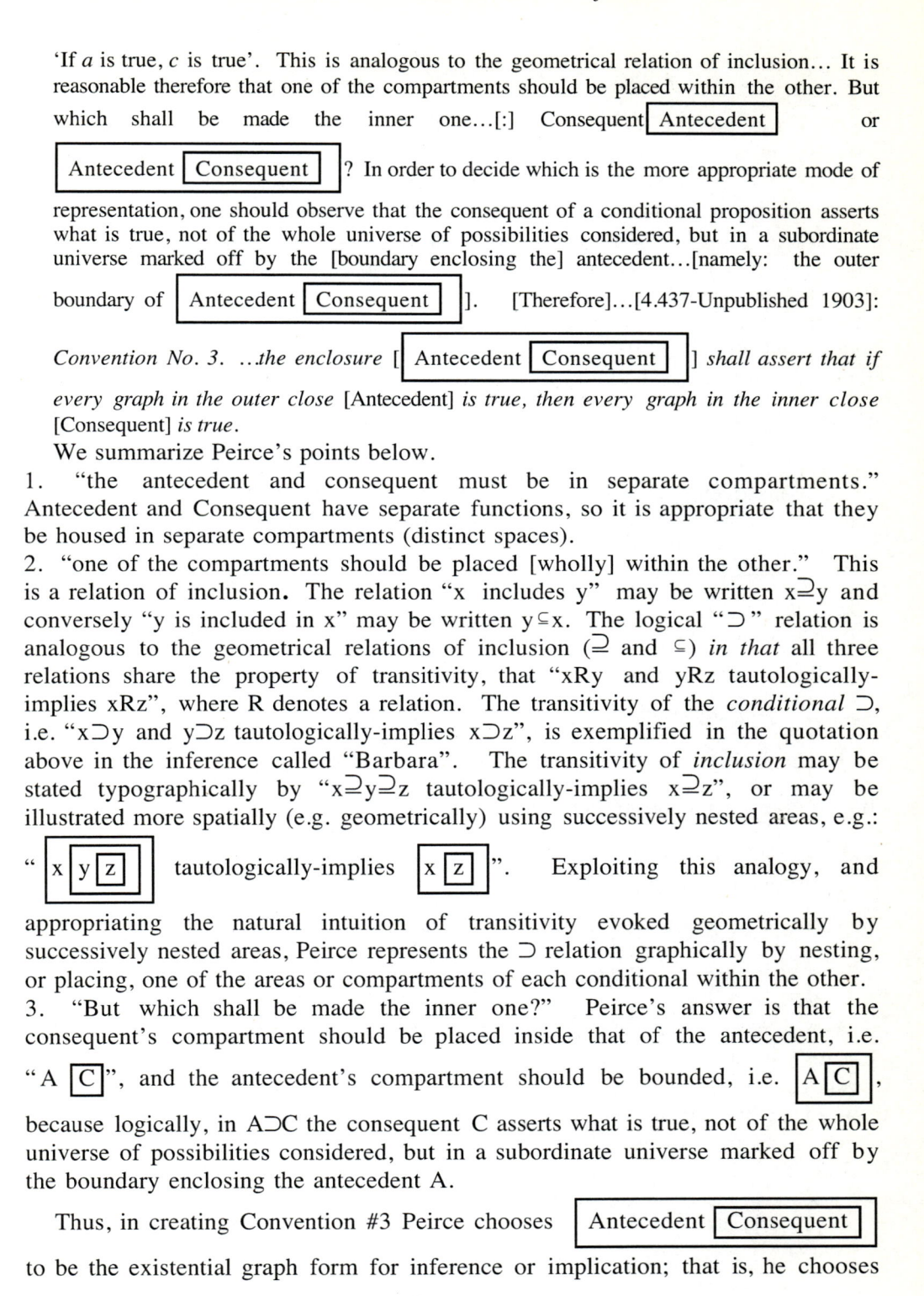

'If *a* is true, *c* is true'. This is analogous to the geometrical relation of inclusion... It is reasonable therefore that one of the compartments should be placed within the other. But which shall be made the inner one...[:] Consequent | Antecedent | or | Antecedent | Consequent | ? In order to decide which is the more appropriate mode of representation, one should observe that the consequent of a conditional proposition asserts what is true, not of the whole universe of possibilities considered, but in a subordinate universe marked off by the [boundary enclosing the] antecedent...[namely: the outer boundary of | Antecedent | Consequent |]. [Therefore]...[4.437-Unpublished 1903]:

Convention No. 3. ...*the enclosure* [| Antecedent | Consequent |] *shall assert that if every graph in the outer close* [Antecedent] *is true, then every graph in the inner close* [Consequent] *is true.*

We summarize Peirce's points below.

1. "the antecedent and consequent must be in separate compartments." Antecedent and Consequent have separate functions, so it is appropriate that they be housed in separate compartments (distinct spaces).

2. "one of the compartments should be placed [wholly] within the other." This is a relation of inclusion. The relation "x includes y" may be written $x \supseteq y$ and conversely "y is included in x" may be written $y \subseteq x$. The logical "⊃" relation is analogous to the geometrical relations of inclusion ($\supseteq$ and $\subseteq$) *in that* all three relations share the property of transitivity, that "xRy and yRz tautologically-implies xRz", where R denotes a relation. The transitivity of the *conditional* ⊃, i.e. "x⊃y and y⊃z tautologically-implies x⊃z", is exemplified in the quotation above in the inference called "Barbara". The transitivity of *inclusion* may be stated typographically by "$x \supseteq y \supseteq z$ tautologically-implies $x \supseteq z$", or may be illustrated more spatially (e.g. geometrically) using successively nested areas, e.g.:

"| x | y | z | | tautologically-implies | x | z | |". Exploiting this analogy, and appropriating the natural intuition of transitivity evoked geometrically by successively nested areas, Peirce represents the ⊃ relation graphically by nesting, or placing, one of the areas or compartments of each conditional within the other.

3. "But which shall be made the inner one?" Peirce's answer is that the consequent's compartment should be placed inside that of the antecedent, i.e. "A | C |", and the antecedent's compartment should be bounded, i.e. | A | C | |, because logically, in A⊃C the consequent C asserts what is true, not of the whole universe of possibilities considered, but in a subordinate universe marked off by the boundary enclosing the antecedent A.

Thus, in creating Convention #3 Peirce chooses | Antecedent | Consequent | to be the existential graph form for inference or implication; that is, he chooses

⊡AC⊡ for the conditional proposition "A⊃C" where A is the Antecedent and C is the Consequent. And a reason for this choice is to show an analogy (both relations being asymmetric and transitive) with the geometrical inclusion relation "A⊇C": i.e. the area in which A stands may be seen to wholly include the area inside the box in which C stands; and to use this geometrical inclusion analogue "⊇" of the logical ⊃ relation as an icon or sign for the logical ⊃ relation, so that "area A geometrically includes area C" denotes the conditional proposition "antecedent A implies consequent C".[16]

Before leaving this long quotation, we point out that Convention No. 3 may be interpreted more generally in order to show feature (j), that of interpreting juxtaposition as conjuction: ∧. Recall "*Convention No. 3. ...the enclosure* [Antecedent ⊡Consequent⊡]] *shall assert that if every graph in the outer close* [i.e. each of the antecedents] *is true, then every graph in the inner close* [i.e. each of the consequents] *is true.*" Here, "*every graph*" means that there may be more than one antecedent and more than one Consequent. For example, for propositions A, B, C, D in existential graphs, Peirce's existential graph "⊡A B ⊡C D⊡⊡" means "(A∧B)⊃(C∧D)", because "A B" means "A∧B" (per the section on feature (j)).

Likewise the (same) expression "⊡A B ⊡C D⊡⊡" in LoF has the (same) meaning "(A∧B)⊃(C∧D)" under interpretation-B.

In conclusion, "A⊃C" is represented as the existential graph "⊡AC⊡", as it also is in LoF notation under interpretation-B.

Feature (b.2): a "cut" as an area/boundary structure

Returning to the level of expressions, and to the first subsection of this section, in Peirce's existential graph ⊡AC⊡ for A⊃C we have a simple example of the way in which each cut creates and defines a *space* inside it *in relation to* the *space* outside it. All the cuts, taken together, create a space/boundary structure such that variables may or may not stand in the spaces. That is, just like LoF boxes (or crosses), existential graph cuts create compartments (spaces) for variables or unmarked space. Cf [4.435-Unpublished 1903]: "the antecedent [A] and consequent [C] must be in separate compartments".

---

[16] By contrast, in Venn diagrams, to be described in a later section, the *converse* inclusion relation, "area A is geometrically *included in* area C", denotes "antecedent A *tautologically*-implies consequent C".

8. *Laws of Form* laws prefigured in existential graphs

Transformations: rules, permissions, conditional principles
In addition to the features of expressions and interpretation noted above, some of LoF's algebraic laws: specifically C1, C5, and C2, are prefigured in laws governing transformations of the alpha part of existential graphs. LoF's algebraic laws are rules governing transformations of LoF expressions. Similarly, Peirce devised principles ("conditional principles") and their applications as rules or "permissions" for the "illative transformation" of existential graphs.[17] Illative transformations are logically valid implications or inferences: that is, tautological-implications.

LoF law C1
Recalling that each enclosure is a negation of its inside, Peirce's Rule 3 in [4.379-published 1911] regarding double-enclosures embodies LoF's algebraic law C1:

$\boxed{\boxed{B}}$ = B.

> [4.379]: "Since two negatives make an affirmative, we have, as Rule 3, that anything can have double enclosures added or taken away, provided there be nothing within one enclosure but outside the other. Thus if B is sometimes true, so that B is written [i.e. the B on the right side of equation C1], Rule 3 permits us to write '[(B)]' [i.e. the $\boxed{\boxed{B}}$ on the left side of equation C1]...[For another example,] Let us make the apodosis [consequent] of a conditional proposition itself a conditional proposition. That is, in (C{D}) let us put for D the [conditional] proposition [A(B)]. We thus have (C{[A(B)]}). But, by Rule 3, this is the same as (CA(B))."

In terms of boxes, (C{[A(B)]}) = $\boxed{C\,\boxed{\boxed{A\boxed{B}}}}$, and the removal of the double-cut {[ ]} around A(B) or $A\boxed{B}$ corresponds to C1: $\boxed{\boxed{E}}$ = E, where E = $A\boxed{B}$. Thus $\boxed{C\,\boxed{\boxed{A\boxed{B}}}}$ becomes $\boxed{CA\boxed{B}}$: i.e. (CA(B)).

LoF law C5
Law C5 or "iteration" is aa=a, or a=aa, and corresponds to

> [4.506-Unpublished c.1903] Rule 2. Called *The rule of Iteration and Deiteration*. ...[any] graph may be iterated...or, being iterated, may be deiterated ...[where] The operation of iteration consists in the insertion of a new replica of a graph of which there is already a replica [a=aa]...[and] The operation of deiteration consists in erasing a replica which might

---

[17] Note: Peirce has different sets of rules at different stages of development and exposition of his existential graphs, so their numbering is dependent upon the particular passage cited. Thus, for example the "Rule 2" of one passage may not be the same rule as the "Rule 2" in another passage.

have illatively resulted from an operation of iteration [aa=a].

LoF law C2 & LoF Theorem 13

C2 or "generation" is ⌈mb⌉b = ⌈m⌉b, or ⌈m⌉b = ⌈mb⌉b, but is easily extended to any number of deeper spaces, if any. For example, given ⌈q⌈p⌈n⌈m⌉⌉⌉⌉b a copy of the b may be inserted into any deeper space: for example, ⌈q⌈p⌈n⌈mb⌉⌉⌉⌉b or ⌈qb⌈pb⌈nb⌈mb⌉⌉⌉⌉b. Likewise, the presence of the b in the outermost space allows any b in an inner space to be erased, so that for example ⌈qb⌈p⌈n⌈m⌉⌉⌉⌉b may be derived from the previous expression. Thus all four expressions (as well as others not shown) are mutually equivalent. Equation C5 (de- and re-iteration), mentioned in the previous subsection, can be included as the "zeroth case" of C2...that is, a copy of the original "b" may be inserted into the space zero depths deeper (for examples, ⌈q⌈p⌈n⌈m⌉⌉⌉⌉b = ⌈q⌈p⌈n⌈m⌉⌉⌉⌉b b, or ⌈m⌉b = ⌈m⌉b b), giving LoF's Theorem 13 [LoF, pp39-40: "The generative process in C2 can be extended to any space not shallower than that in which the generated variable first appears."

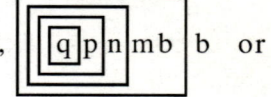

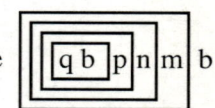

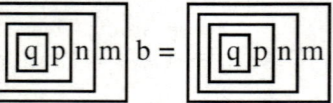

Theorem 13 turns out to be the same as

[4.492-Unpublished c.1903]: "Rule 2. Any graph may be iterated within the same or additional [i.e. deeper] seps [cuts, or boxes], or if iterated, a replica may be erased, if the erasure leaves another outside the same or additional [i.e. shallower] seps."

The two parts about "additional seps" cover C2 and its extension to deeper boxes. If the parts about additional seps are ignored, Rule 2 covers (corresponds to) LoF's C5.

9. Differences between steps in existential graphs and *Laws of Form*

We have shown that LoF's equations C1, C5, and C2 corrrespond with some of Peirce's illative transformation rules, but other LoF equations do not seem to corrrespond to existential graphs rules: for example J2, C4, C6, C7, C8, C9 (see [LoF, pages 139-41]...J2 will be shown later). Likewise, some of Peirce's illative rules seem to have no counterpart in LoF equations.

In order to better understand how some of LoF's algebraic laws corrrespond with Peirce's "illative transformation" rules, and why some corrrespond and some do not, we point out three ways in which the steps of transformation for existential graphs and LoF differ: 1. rules in words *vs.* in templates (or non-iconic *vs.* iconic); 2. analytic *vs.* calculational steps; and 3. steps of tautological-implication *vs.* equality.

Words vs. *templates: non-iconic* vs. *iconic steps*
One difference is that Peirce's rules (in contrast to his iconic expressions) are presented in words which must be processed intellectually into an "abstract" understanding, whereas LoF's algebraic laws are (mostly) presented in "concrete" iconic templates. An example of the former non-iconic abstractness is Peirce's
> [4.415-published 1903: Permission No. 6. The reverse of any transformation that would be permissible on the sheet of assertion is permissible on the area of any cut that is upon the sheet of assertion.

Examples of LoF's concrete iconic, templates are C1: $\overline{\overline{a}}$ = a and C5: aa = a.

Analytic vs. *calculational steps*
Peirce designed the steps of transformation of his existential graphs to be as analytically discerning as possible—that is, to use as many steps as possible, as the following quotation shows.
> [4.424-Unpublished c.1903]: ...this system [of existential graphs] is not intended to serve as a calculus, or apparatus by which conclusions can be reached and problems solved with greater facility than than by more familiar systems of expression...The principal desideratum of a calculus is that is should be able to pass with security at one bound over a series of difficult inferential steps. ...But in my algebras and graphs, far from anything of that sort being attempted, the whole effort has been to dissect the operations of inference into as many distinct steps as possible.

On the contrary, Spencer-Brown designed LoF to be a calculus with as few steps (calculations) as possible.

Not only does Peirce want as many steps as possible, he wants the steps to be as elementary as possible:
> [4.429-Unpublished c.1903]: Our purpose [for existential graphs]... is to study the workings of neccessary inference. What we want, in order to do this, is a method of representing diagrammatically any possible set of premises, this diagram to be such that we can observe the transformation of these premises into the conclusion by a series of steps of the utmost possible simplicity. [And in 4.377-published 1911]: we are to recognize no transformation as elementary except writing down and erasing [alternately].

In light of their different purposes, we can see why, for example, J2: $\overline{\overline{a}\,\overline{b}}\,c$ = $\overline{\overline{ac}\,\overline{bc}}$ (a LoF equation for factoring and distributivity) is not included in the rules of transformations for existential graphs: namely, it conflates steps that Peirce wants to keep separate. J2 both erases and inserts, or erases and writes down, in the same step. For example, in the direction of distribution the term *c* is erased from the outermost space and inserted into the innermost spaces. For Peirce, in order to ensure the utmost simplicity while at the same time using as many distinct steps as possible, writing down must be one step and erasing must be another. Therefore,

J2 is not used in the existential graph system. (By the way, if in J2 writing down and erasing were done in seprate steps the equality would no longer hold.)

The same reasoning applies to LoF equations C4, C6, C7, C8, and C9.

A-implies-B vs. *A=B*

In existential graphs Peirce is concerned with "illative transformations", that is, logically valid implications or inferences (technically they are tautological-implications) whose validity is claimed in one direction only—that is, the illation "A tautologically-implies B" makes no claim that "B tautologically-implies A";

By contrast, the (algebraic) transformations in LoF are exclusively *equalities* "A=B", and are valid in both directions: "A tautologically-implies B *and also* B tautologically-implies A".[18]

For an example of a valid illative tranformation (tautologically-implication) that is only valid in the one direction, and can therefore not be an equation, we may use

> [4.492-unpublished 1903]: *Rule* 1 [first part only]. Within an even number (including none) of seps [cuts or boxes], any graph may be erased...

If it were an equality it would say *"may be erased or inserted"*. But such an insertion is not valid, so it cannot be an equality, and so we can not expect to find a corresponding LoF equation.[19]

10. Euler-Venn diagrams

Although so far in this paper we have been mainly comparing two diagrammatic systems for logical inference, namely existential graphs and LoF (with minor reference to Peirce's first attempt, his entitative graphs), Peirce himself was well aware [4.350-4.356-c.1903] of a system of diagrams (1772) by Euler, and of its revival by Venn (1881 per [4.357]), in what are now known as Venn diagrams. These diagrams provide a way of illustrating logical possibilities and logical relationships and of drawing inferences from these. In this type of system, a circle is assigned to each property or class. The interior of the circle has the assigned property, whereas the space outside the circle does not. In terms of classes, whatever is in the circle is in the class, whatever is outside the circle is is not in the class. To illustrate how this works we simplify by just considering two properties, A and B. For convenience we will use boxes instead of circles, so we have " $\boxed{A}$ "

---

[18] Interestingly, in Appendix 2 of LoF, tautological-implications—i.e., one-way only—are dealt with using equalities. Cf [LoF, bottom p 118].

[19] Technical note: The validity of the zeroth case of [4.492]'s Rule 1 first part, that any graph (call it "g") not in a box may be erased, follows from the logical fact that "For every possible value of g and h, $g \wedge h \supset h$ is true and $h \wedge g \supset h$ is true" [Carnap, page 27, T8-2 E.b.(1) and (2)]...That this extends to any even number of boxes may easily be seen by the fact that each successive enclosing box negates its inside, together with the fact that an even number of compounded negations cancel each other. But the converse, that any graph not in a box may be *inserted*, may be seen to be *invalid* from the fact that "h does *not* imply $g \wedge h$", which is demonstrated by the case in which g=false and h=true.

and "[B]". Boxes may:

- include others, as in [A[B]], where [A] includes [B];
- or not, as in "[A] [B]";
- or may (unlike existential graphs and LoF) partially overlap and intersect, as in [A[⊓]B];
- or not, as in the first two examples above;

The Venn diagrams above indicate that "all B are A", "no A is a B", and "only some A are also B", respectively.

Notice that the third example, [A[⊓]B], in which the boxes intersect, allows for all four logical possibilities in terms of four mutually exclusive areas: neither A nor B (outside the boxes); both A and B (in the intersecting area inside both boxes; A and not B (inside box A but outside box B); and finally, B and not A (vice versa). In the other two examples one of these four logical possibilities and areas is missing. The first example, [A[B]], is missing the logical possibility of a B that is not an A.

An inference that can be drawn from the diagram "[A] [B]" is the converse of "no A is a B", namely that "no B is an A".

11. Comparing Euler-Venn diagrams, existential graphs, and *Laws of Form*

There are several related threads to weave that will shed light on the three systems—existential graphs, LoF, and Euler-Venn diagrams—by comparing and contrasting them with regard to following interrelated topics: 1. representing logical implication and also transitivity in terms of inclusion ($\supseteq$ or $\subseteq$) diagrammatically in terms of circles or boxes; 2. the "point-class interpretation" of certain diagrams; and 3. the application of the above topics to the inference called "Barbara". It will turn out that interpretation-B of LoF and the alpha part of existential graphs are relatively compatible with each other in their expressions and their interpretation and in transformational steps, and both systems are incompatible with Venn diagrams and their interpretation and steps.

Comparison of the three systems

In this subsection we will apply topics 1. and 2. above to the following box-diagrams: [A[C]], [A]C], and [A]C. (You may have to look carefully to distinguish the first two diagrams ...the inner box has a different content.) We will correlate these three diagrams with the three systems[20] and with the three forms related to logical implication—either the conditional proposition A⊃C, or else the universal statement "all A are C", or else the tautological-implication " A tautologically-implies C". Some of these will also correlate with a direction of inclusion (either ⊇ or ⊆) in terms of boxes.

The first set of correlations is as follows. Existential graphs and LoF interpretation-B both use $\boxed{A\boxed{C}}$ for the *conditional* A⊃C. LoF interpretation-A and LoF interpretation-C both use $\boxed{A}$ C for the *conditional* A⊃C, but in LoF interpretation-C $\boxed{A}$ C is also the *universal* proposition "all A are C" *taken as an hypothesis*. Venn diagrams use $\boxed{A\boxed{C}}$ to mean "A *tautologically*-implies C " and also the *universal* statement "all A are C" *taken as a fact*.

Now we give reasons which motivate or justify these correlations. We pick up the thread from the section on Feature (c.2), where Peirce denoted the conditional proposition "A⊃C" by the existential graph $\boxed{A\boxed{C}}$, in which the Antecedent A lies outside the Consequent C, and where $\boxed{A\boxed{C}}$ is an icon for the geometrical inclusion relation "A⊇C". Recall the motivation that Peirce gave in [4.435-Unpublished 1903] for his choice of $\boxed{A\boxed{C}}$ rather than $\boxed{A}$ C to represent the *conditional* implication "if A is true, C is true":

> ...the consequent of a conditional proposition asserts what is true, not of the whole universe of possibilities considered [as would $\boxed{\text{Antecedent}}$ Consequent], but in a subordinate universe marked off by the [boundary enclosing the] antecedent [namely: the outer boundary of $\boxed{\text{Antecedent}\boxed{\text{Consequent}}}$].

His choice is thus made so as to be consistent with a particular diagrammatic interpretation of the logical concept of "truth within a subordinate universe".[21]

By contrast, in Euler-Venn diagrams the converse is the case—the Consequent

---

[20] ...or five systems if we count LoF interpretations-A, -B, and -C separately.

[21] But we should note that this choice was not independent of others. In light of the close correspondences between the alpha part of his existential graphs and interpretations-B in LoF, Peirce's choice of A⊃C as $\boxed{A\boxed{C}}$ rather than as $\boxed{A}$ C may be seen to be forced by his interpreting the sheet of assertion as "true" rather than "false" or by his interpreting AB as A∧B, i.e. as asserting A and B conjunctively rather than disjunctively as A∨B. This forcing is because these three *former* choices taken together are mutually consistent, as are the three *latter* ("rather than") choices taken together, but combinations that mix the two sets are not.

C lies outside the Antecedent A—in ⌐C⌐A⌐, denoting "C⊇A", which is equivalent to "A⊆C" or "⌐A⌐C⌐". Euler-Venn diagrams have a different motivation for choosing "⌐A⌐C⌐" rather than "⌐A⌐C⌐⌐" to represent the *universal* statement-of-fact "all A are C" based on a *different* diagrammatic interpretation of a "subordinate universe", that of a subclass [4.350, unpublished, c. 1903]. Thus "all A are C" can be restated as "the class of A's is entirely contained or included in the class of C's". Or, "if some x is in class A, then that x is also in class C". In modern set theory, this is essentially the definition of subclass: "A⊆C" means that, given any individual x and classes A and C, "(x∈A) ⊃ (x∈C)" is a tautology, where "x∈A" means "x is in A". Interpreted spatially (or "geometrically", but without the irrelevant metric) we may call this the *"point-class interpretation"*, where the various possible individual "x's" are all of the points in the space and where outer compartments (classes) include inner compartments (subclasses).

In a loose sense we might say that LoF accommodates these opposite cases (existential graphs and Venn diagrams) as its interpretations-B on the one hand and -A and -C on the other, respectively, but there are two caveats. One is that in interpretations -A and -C "A⊃C" becomes "⌐A⌐C"[22],[23]—with no boundary around the C, rather than "⌐A⌐C⌐". Another is that in interpretations-A and -C "⌐A⌐C" denotes the conditional or *hypothetical* "A⊃C", which is not identical with "A⊆C" or "A tautologically-implies C" or the *universal* statement "all A are C", all *taken as fact*.

In LoF interpretation-C the *universal* statement "all A are C" is first cast into a conditional proposition of the form (x∈A) ⊃ (x∈C), and thence successfully conflated to simply A⊃C, which is represented, just as in LoF interpretation-A, as ⌐A⌐C, i.e. ⌐A⌐C, per our LoF primer and the section on Feature (c.1), and per [LoF pages 114, 119].

The point-class interpretation works for the Venn diagram ⌐A⌐C⌐ but *not* for

---

[22] LoF's "⌐A⌐C" under interpretation-A is analogous to "⌐h⌐—d" in Peirce's earlier entitative graphs [3.475-Published 1897].

[23] One reason that "⌐A⌐C" is used rather than "⌐A⌐C⌐" to represent "A⊃C" in interpretation-A has to do with the fact that the former, not the latter, bears the logical relation of 'duality' [Cu p294] to the interpretation-B form ⌐A⌐C⌐.

the existential graph, which has the opposite form $\boxed{A\,\boxed{C}}$; and it *appears* to apply to the LoF expression $\boxed{A}$ C under interpretations-C and -A, but does not, as shown by the fact, pointed out in the "second difference" below, that the point-class interpretation does not extend to further nestings.

We conclude this subsection by noting the following three sorts of differences between Venn diagrams and the other systems.

First, in LoF interpretations-C and -A the expression $\boxed{A}$ C s the mere conditional A⊃C, which may or may not be true; whereas the Venn diagram $\boxed{A\,\boxed{C}}$ asserts that class A *is* a subclass of class C, which corresponds with asserting that the conditional A⊃C is a tautology.[24]

Second, the point-class interpretation may be extended from $\boxed{A\,\boxed{C}}$ to $\boxed{A\,\boxed{B\,\boxed{C}}}$ in Venn diagrams (see the subsection on "Barbara" in Venn diagrams), but not in existential graphs, nor in LoF from $\boxed{A}$ C to $\boxed{A\,\boxed{B}}$ C. This has to do with the first difference above, beween: the conditional A⊃C, which is a value to be calculated from the operation ⊃; and the conditional A⊃C-that-is-a-tautology, which is an ordering A⊆C of values A and C (see the section on logic and recall the analogy between A⊆C and "1≤2") The difference is graphically apparent in that for LoF and existential graphs $\boxed{\boxed{A}}$ = A, per the section on C1, and embodying the fact of logic that "not-not-A = A"; whereas for the relation ⊆ and for Venn diagrams $\boxed{\boxed{A}}$ is not equivalent to A, as each boundary represents a relation (or part of a relation) of ⊆, and not a logical negation operation.

Third, Venn diagrams differ from both LoF and existential graphs in their use of boundaries, variables, and empty space. In Venn diagrams a variable signifies a class, which creates a subordinate universe, which is diagrammed as an area of (normally unmarked) points enclosed in a boundary. If the class is empty, it has no points, it thus cannot be represented at all in the diagram: the blank space (the area), the variable, and the enclosing boundary all must disappear. By contrast, in both LoF expressions and existential graphs, blank space, variables, and

---

[24] Note: if, instead, the proposition "A is a subclass of C" were taken as a conditional that was *not* a tautology, it's Venn diagram would be two boxes that overlap and intersect: $\boxed{A\,\boxed{}\,B}$

(a possibility denied in LoF and in the alpha part of existential graphs, where boxes or cuts may not overlap or intersect), allowing for some A that was *not* C.

boundaries all are relatively independent of one another: a variable may or may not stand in an area, but it does not "fill" the area (these variables are more like points than areas), nor is it dependent on any boundary that may or may not enclose it. And if a variable has value "unmarked", that variable is still an expression, and can stand in the space, and does not require that any boundary enclosing it be erased.

"Barbara" in Venn diagrams
The three diagrammatic systems—existential graphs, LoF, and Venn diagrams may be further contrasted using the inference called "Barbara", an example of which is: "If all men are animals and if all animals are mortal, then all men are mortal". Recalling the definition of transitivity (from the previous section on Feature (c.2)), we see that Barbara's logical structure is that of the transitivity of the relation "all A are C". Per the previous subsection, this relation can be restated more technically as the follows: "Given any individual x and classes A and C, (x∈A)⊃(x∈C) is always true, i.e. is a tautology, where x∈A means that x is in A." But this is precisely the inclusion relation A⊆C, i.e. A is a subclass of class C. So Barbara's structure may be seen in the transitivity of the inclusion relation ⊆: i.e. " A⊆B and B⊆C tautologically-implies A⊆C".

What, then, is the Venn diagram for the inference called "Barbara"? Under the point-class interpretation in the previous subsection the inclusion relation A⊆C may be represented spatially by its Euler-Venn diagram [A[C]], where each class is represented by the *entire* area within the boundary enclosing its name (A, C, etc). So Barbara "[A[B]] and [B[C]] tautologically-implies [A[C]]", where the first two diagrams are called the premises and the last is called the conclusion. So far the Venn diagrams are not very illuminating in suggesting transitivity. What Euler did back in the 1700's was to draw the following nested figure: [A[B]C], which interprets the "and" in Barbara diagrammatically by replacing the [B] part of the second premise, [B[C]], by the first premise, [A[B]]. It thereby graphically shows the relation " A⊆B and B⊆C" as the linear ordering " A⊆B⊆C". Now this *is* illuminating, for now the transitivity of ⊆ is evident, based, for example, on our understanding of the transitivity of the relation "is less than (<)" in the linear ordering "1<2<3". That is, just as "1<2<3 tautologically-implies 1<3", likewise " A⊆B⊆C tautologically-implies A⊆C", and " [A[B]C] tautologically-implies [A[C]]". That is, because each class (A, C, etc.) is represented by the *entire* area within the boundary enclosing its name (A, C, etc.), any point x that is in area A is also in area C. As Peirce describes the application of Euler's diagram to "Barbara" in [4.350, unpublished, c. 1903]: "We see, then, that whatever is enclosed in the circle [surrounding A] is enclosed in the circle [surrounding C]; that is, all [A] are [C]." So the contribution of the Euler-Venn diagram $\boxed{\boxed{A}\boxed{B}C}$ to the inference "Barbara" is that, by combining the premises into an ordering (here, a nesting of boxes), it suggests the conclusion $\boxed{\boxed{A}C}$, under the point-class interpretation.

"Barbara" in existential graphs compared with LoF
We mentioned at the end of the section on LoF that the inference "Barbara" can be transcribed and justified using LoF interpretation-C. But it is LoF interpretation-B which corresponds with the alpha part of existential graphs, and interpretation-C follows interpretation-A, not B. Moreover interpretation-C involves the possibility of what in logic are called existential statements, a possibility inconsistent both with LoF interpretations -B and -A. So interpretation-C involves a level of complexity beyond the scope of this paper, and will not be further discussed.

Peirce applies his existential graphs to "Barbara" in [4.571, text & Figs. 211-8], using both Alpha and Beta parts, which we reproduce below except that we simplify by deleting the Beta parts. When this is done, the remaining Alpha parts are interpretable as LoF expressions under interpretation-B. In both existential graphs and LoF interpretation-B "Barbara" would be stated as " $\boxed{\boxed{A}B}$ and $\boxed{\boxed{B}C}$ tautologically-implies $\boxed{\boxed{A}C}$ " (here we are mixing English with the graphic expressions). The first figure, 211, begins by transcribing the premises " $\boxed{\boxed{A}B}$ and $\boxed{\boxed{B}C}$ ". Because the logical relation "and" (logical conjunction) is represented in these systems by juxtaposition, the premises become simply " $\boxed{\boxed{A}B}$ $\boxed{\boxed{B}C}$ ". Each successive figure is obtainable as a logically valid inference. To the right of each figure, we adduce a logical justification, either from LoF or from Peirce's rules:

	Existential graph	Justification and explanation
Fig. 211:	$\boxed{\boxed{A}B}$ $\boxed{\boxed{B}C}$	; Given, premisses (antecedents)
Fig. 212:	$\boxed{\boxed{A}\boxed{B\boxed{C}}}$ $\boxed{\boxed{B}C}$	; extended-generation (C2) of $\boxed{\boxed{B}C}$

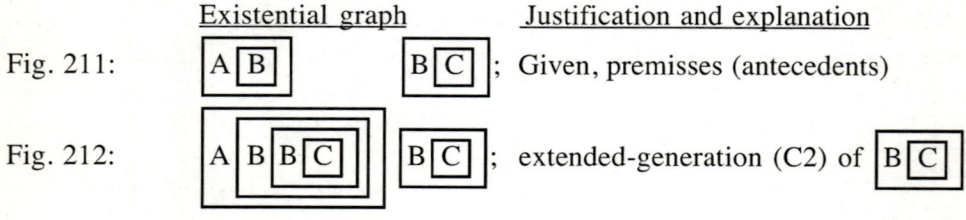

Fig. 213:   A B [B [C]] ;   Rule 1a (erasure), on [B [C]]

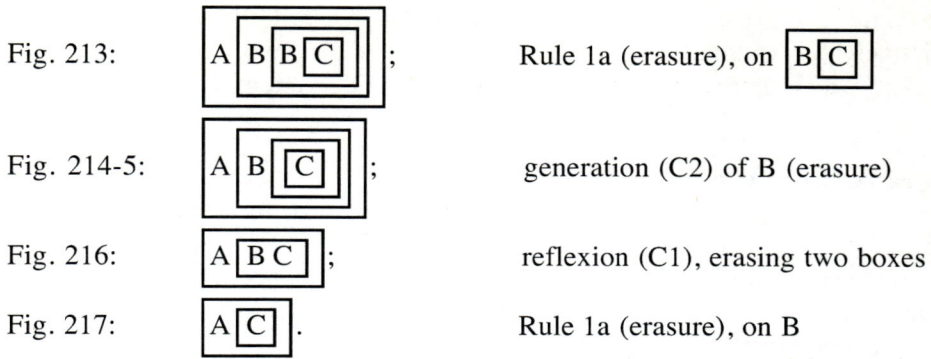

Fig. 214-5: A B [C] ;   generation (C2) of B (erasure)

Fig. 216:   A [B C] ;   reflexion (C1), erasing two boxes

Fig. 217:   A [C] .   Rule 1a (erasure), on B

All but the second and last steps can be done in LoF using equations C1 and C2 (which have equivalents in Peirce's Rules, as seen in the section "LoF equations C1, C5, & C2 prefigured in illative rules of existential graphs")…Recall that C1 and C2 each have two directions, one that erases and one that "adds to". The second and last steps steps may be done *via* Peirce's Rule 1a of [4.492-unpublished 1903] a justification of which was cited in the section "3. 'A tautologically-implies B' *vs*. 'A=B' ". Alternatively, the validity of the two steps could be proved *via* the method in [LoF, Appendix 2, pages 115-118].

Comparison of the three systems in terms of "Barbara"

Comparing the three systems in terms of "Barbara", recall that in the Euler-Venn diagram: 1. the antecedent lies *inside* the consequent and the point-class interpretation *holds*; 2. the premises or antecedent in "Barbara" is " [A B] and [B C] ", which becomes, *not* " [A B] [B C] ", but rather [A B C] ; and 3. the conclusion or consequent [A C] follows in one step (via the transitivity of inclusion). By contrast, in both existential graphs and interpretation-B of LoF: 1. the antecedent lies *outside* the consequent and the point-class interpretation does *not* hold; 2. the premises or antecedent in "Barbara" is " [A B] and [B C] ", which becomes " [A B] [B C] " (the word "and" joining the premises translates as simple juxtaposition) and cannot be nested; and 3. in existential graphs the conclusion [A C] takes several steps.[25]

---

[25] These steps may involve nesting, but none of the steps correspond to the "Venn-like" graph

To conclude, interpretation-B of LoF and the alpha part of existential graphs are relatively compatible with each other in their expressions and their interpretation and in transformational steps, and both systems are incompatible with Venn diagrams and its steps.

12. Did Peirce influence *Laws of Form*?

This section is less rigorous than the others, but I include it for reasons of general interest.

I do not know definitively, but I suspect that Peirce did not influence Spencer-Brown in creating his *Laws of Form* system. My major (circumstantial) evidence is that Peirce is cited just once in *Laws of Form*:

> To the best of my knowledge, Peirce was the only previous author to recognize, as such, what I call position [$\overline{a}|a|$ = unmarked]. He called it erasure[13], thus again drawing attention to only one direction of application. [LoF p136];

and this citation of Peirce's "erasure" appears to the author, both at first glance and in light of the details, to be more retrospective than causal. At first glance, if that is the extent of Spencer-Brown's crediting of Peirce as a predecessor, then it appears negligable. Examining the details (below), my research shows that the "erasure" cited does not even refer to Peirce's existential graphs, but rather to an obscure typographical notation, thus making any connection even more unlikely.

Here are the details. The erasure that Spencer-Brown refers to is " $\overline{a}|a|$ tautologically-implies unmarked"; that is, $\overline{a}|a|$ may be erased. Although the reference for this citation is to Peirce's entire volume IV (i.e. [4]), and thus lacks specificity, the citation presumably refers, not to Peirce's mentions of erasure in his existential graphs, but rather to earlier work [4.12-4.20] c. 1880 in a quite different typographical notation. For when Peirce mentions erasure in the alpha part of his existential graphs, he means a logically valid inference of the following sort, [4.492-unpublished 1903]: *"Rule 1a.*. Within an even number (including none) of seps [cuts or boxes], any graph may be erased..."[26] This erasure, under Peirce's Rule 1a, of *any* graph *g*, restricted to an *even* number of boxes, is logically quite different from the erasure of the more proscribed graph $\overline{a}|a|$

$A|B|C||$, which of course is not a Venn diagram because the ordering by depth of the variables is reversed regarding consideration 1 in this paragraph.

[26] We encountered this rule in our sections "A-implies-B *vs.* A=B" and " "Barbara" in existential graphs compared with LoF".

under the rule " $\overline{a|a}$ tautologically-implies unmarked", which may be erased within *any* number of boxes (including none), regardless of whether even or odd. Examining Peirce's use of erasure in his earlier work [4.12-4.20] c. 1880 one finds an obscure typographical notation, and it does not seem to this author to be a likely inspiration for Spencer-Brown's position [$\overline{a|a}$ = unmarked], much less for LoF notation. Thus Spencer-Brown's single citation above of Peirce appears to be more retrospective than causal.

Another consideration, again merely circumstantial evidence, is that while Peirce specifically wanted a graphic notation due to his conviction that all reasoning is, in its essence, iconic, my sense is that Spencer-Brown was not similarly pre-disposed to his graphic notation with its peculiar features, but rather, that the boxes and unmarked value were unpremeditated discoveries which "dawned upon him" during his inquiry.

So I suspect that Peirce did *not* influence Spencer-Brown in creating his *Laws of Form* system.

Appendix 1. A *Laws of Form*-like modification of Peirce's notation
Departing from Peirce's graphs to other algebraic notations, I propose below a modification of Peirce's use of a horizontal line to give it LoF-like properties. When this is done the line converts any operation to its dual operation.

Peirce often uses a horizontal line (vinculum or obelus [4.391] and [3.494]) over a *proposition* symbol (a letter or variable) to negate it (to interconvert its value between "true" and "false"). Thus the negative of proposition y is denoted by $\overline{y}$, and, for example, the negative of the proposition x⊃y, involving the operation ⊃, is $\overline{x \supset y}$.

But it turns out there could be several possible meanings by which a horizontal line (vinculum or obelus) over an operation *symbol*, such as $\overline{\supset}$, might be defined. In [3.386-3.387] Peirce defines a horizontal line (vinculum or obelus) over his conditional operation symbol (which I here replace with the modern symbol ⊃) to mean negating the result of that operation: $x \overline{\supset} y = \overline{x \supset y}$, or with different formatting, $x\overline{\supset}y = \overline{x \supset y}$. (The symbol $\overline{\supset}$ is not to be confused with the symbols for inclusion, ⊇ and ⊆.) Yet he also defines another kind of negative of an operation, the "internal negative operations" [4.295-unpublished 1902], namely the pairs (using modern names) conjunction and disjunction, NAND and NOR, and others. The modern name for "internal negative operations" is "dual operations".

I propose that, instead of a vinculum above an operation symbol meaning negating the result of that operation, it denote the *dual* operation, for then the strong (De Morgan) duality that holds in logic [Curry pp293-294] ensures the following notational nicety: that any vinculum over the result of an operation $*$ is equivalent and thus interconvertible with separate vinculi over each symbolic component, that is, over the operation and both operands: $\overline{x*y}$ $\overline{x}\;\overline{*}\;\overline{y}$. In the case of the conditional this is: $\overline{x \supset y} = \overline{x}\;\overline{\supset}\;\overline{y}$. The De Morgan laws in this notation become $\overline{x \vee y} = \overline{x}\;\overline{\vee}\;\overline{y}$ and $\overline{x \wedge y} = \overline{x}\;\overline{\wedge}\;\overline{y}$, where $\overline{\wedge} = \vee$ and $\overline{\vee} = \wedge$. This equivalence holds for all five possible dual pairs of logical operations[27], and holds regardless of whether pre-fix (=reverse Polish), post-fix, or in-fix (the usual) typographical notation is used.

This modification of Peirce's vinculum notation has LoF-like properties, analogous to the LoF equations I1: ⌐=⌐ ⌐=⌐⌐ (called "number") and C1: $\overline{\overline{x}} = x$. The analogy with I1 is that in both cases a single horizontal line (or cross or box), as a unary operation whose scope is defined by what stands under it, can be divided into separate multiple lines (or crosses or boxes). If we write just the top line of LoF crosses or boxes, a single cross could divide, per I1, in three like ¯ = ¯ ¯, in similarity to the vinculum dividing in three in $\overline{x*y} = \overline{x}\;\overline{*}\;\overline{y}$. As for C1, since both uses of the vinculum in this proposed interpretation, namely negation-of-value (or negation-of-proposition) and dual-of-operation, are self-inverse-operators, in every case we have that a vinculum over a vinculum cancels, namely $\overline{\overline{x}} = x$, in precise analogy to the form of, and with the same meaning as, LoF's C1: $\overline{\overline{x}} = x$. Thus, for example, $\overline{\overline{x*y}} = x*y$. Another example, this one using both analogies, is $\overline{\overline{x}\;\overline{*}\;\overline{y}} = \overline{\overline{x}}\;\overline{\overline{*}}\;\overline{\overline{y}} = x * y$.

Appendix 2. "A notation unifying wholes and parts..."

A notation unifying wholes and parts: G. Spencer-Brown's *Laws of Form*[28]
Jack Engstrom

G. Spencer-Brown's book *Laws of Form* (1969), referred to here as "LoF", offers a unique contribution both to current mathematics and other disciplines which might model relationships involving wholeness and parts. That contribution is a vision in which forms and relations arise from wholeness and can disappear

---

[27] More precisely, of *binary* logical operations: that is, of operations which involve two terms. All of the operations in this paper are binary operations.
[28] February 1996. This essay is a condensation of Chapter III, "*Laws of Form*, Wholeness, Mathematics, and Metaphysics", of Jack Engstrom's Master's thesis in mathematics, "Natural numbers and finite sets derived from G. Spencer-Brown's *Laws of Form*", Maharishi University of Management, Fairfield, Iowa, June 1994, Revised December 1994, 160 pages, unpublished.

back into that wholeness ("voidability of relations" LoF, p104). To be whole means to be undivided, complete, hale. Mathematics and other notations deal with relations: generally relations between different parts. In the deepest analysis, parts cannot be taken as given; parts must ultimately be generated by dividing wholeness.

Based on
(0) wholeness,
the essence of the vision and notation of LoF is two-fold:
(1) form—including the notation itself—arises *at all* as the *relation* that results from taking apart wholeness in the act of distinguishing — thus form arises out of formlessness; and
(2) relations are voidable, and thus forms (parts) can be *transcended* back into the wholeness from which they were originally created.
We will explore these three themes, (0), (1), and (2), in this essay.

How does LoF allow form to be transcended back into wholeness? LoF *starts with unsevered wholeness*, represented by unmarked space, say a blank sheet of paper. It isn't that an umarked space *is* wholeness, but rather that it can, in its most profound interpretation, represent wholeness. This interpretation perhaps requires some explanation. The *only* arithmetical marks (forms) called forth into existence are those peculiar marks, the LoF "cross", that are seen (as an abbreviation of a box) to cleave the spaces in which they stand into distinct inside and outside spaces. Thus the LoF cross—the only arithmetical symbol in the LoF notation—stands apart from commonly used mathematical and linguistic symbols because it derives both its form and its meaning not from highly arbitrary conventions (examples: the Greek letter epsilon stands for set membership; the letter "x" stands for a variable) but from its "topological" *relation* to the *spaces* it distinguishes. An unmarked space is unsevered and therefore *whole*, whereas each LoF cross cleaves that wholeness into two *parts*: an inside space and an outside space, the relation between the two parts being indicated visually by the form separating them. An unmarked space *has no parts* and is thus unsevered and is therefore *whole*. It is in *this* sense that an unmarked space can naturally represent wholeness. The second axiom of LoF allows a voiding of two nested crosses (forms) *back to the original unsevered wholeness*.

Regarding themes (0) and (2) together, we see that the transcendence that characterizes wholeness in LoF is a *transcendence of form, of relation*. In contrast to approaches (such as set theory in mathematics) which try to get wholeness by a web of ever-larger and more complex *relations*, LoF's second axiom allows a transcendence of form back to wholeness by *voiding form, voiding relations*. We propose that wholeness is *not* a universe, *not* a creation, *not* a class, *not* a relation or web of relations but rather the *uncreated* foundation *out of which* these are created.

In the context of (0) and (2), we now explore (1), the thesis that form arises *at all* as the *relation* that results from taking apart wholeness in the act of distinguishing: that manifest form ultimately *arises out of* unmanifest, formless wholeness. Naïvely, one tends to take the basic "furniture" of one's universe — e.g., chairs in philosophy, rocks in physics, numbers in mathematics (or nowadays, sets) — as given. Modern mathematics analyzes numbers and most everything else mathematical into sets which obey the postulates of Zermelo-Fraenkel set theory, which in turn are formalized via the "universal" language of first-order predicate logic plus the membership relation into a typographical string of symbols built up according to certain rules of formation. The symbols themselves are *highly arbitrary* conventions with no intrinsic ability to evoke or connote their meanings.

So how can we have a relatively non-arbitrary fundamental symbol arise from dividing wholeness? LoF accomplishes this by starting with an unsevered space, representing wholeness, and then severing it with a distinction, say a box abbreviated to a LoF cross, the form of which now distinguishes *two* spaces and at the same time expresses visually *relations* between them: they are contiguous but distinct; the space inside the cross is bounded by the outline of the cross; the space outside is "unbounded". So the LoF cross is a prime example of (1) notational form arising out of the formless as the *relation* that results from taking apart wholeness in the act of distinguishing.

In conclusion, we claim that LoF's unmarked state and cross, in their deepest interpretations, mathematically model uncreated wholeness and the seed of creation, respectively, which sprouts into the entire universe of forms. Wholeness "waters the root" of the tree of manifest creation which has sprouted from this single seed-form.

Appendix 3. *Laws of Form* and the Mystical Void

Laws of Form and the Mystical Void[29]
Jack Engstrom

The concept of the mystical void is a recurring theme in spirituality. It can be characterized as pure consciousness -- a single, undifferentiated state, which contains all possibilities and from which all distinctions emerge.(1) So if we are to understand the nature of reality in terms of the mystical viewpoint, we must start from the idea of the undifferentiated void, to which a concept of distinction can be added. How then can this idea be adapted to science, in which the basic laws are expressed in terms of distinctions?

Laws of Form, a mathematical system developed by G. Spencer-Brown(2), is based on the concepts of the void and distinction.* Obviously, the concept of a void which underlies all distinction is a profound one, and to fully understand it, we should first understand how such seemingly different ideas as wholeness and the void can be related. We will then see that using Laws of Form to undergird

---

[29] Originally published in the *Newsletter of the Center for Unified Science Within Consciousness*, Vol. 1, No. 1, Spring 1996, and at http://www.swcp.com/swc/Spring96/9601engst.htm. Republished with permission of Henry Swift.

modern science can provide a way to relate current science to spiritual ideas.

Wholeness. The concept of wholeness is not easy to define. *Whole* means hale, healed, unharmed, undivided, unsevered, complete. Wholeness is certainly not a thing, not a part. It might be thought that wholeness is a unity of parts, but even a unity of parts is a distinction. Wholeness is not a relation nor a web of relations. Nor is wholeness a creation. Rather, it is the uncreated foundation out of which both parts and relations are created. Ultimately, wholeness and the mystical void are the same.

Then how might this wholeness/void be represented? If it is the uncreated foundation of everything, it must be a no-thing. Abstractly, no distinctions or separations can exist. Thus in Laws of Form the symbol for the void is a blank space on a page -- not any kind of marking at all, just blank space.

Generating form by dividing wholeness. In order for a form to arise out of unmarked space a distinction must be made. In other words, wholeness must be taken apart by the act of making a distinction. This act can be symbolized by drawing a box in the blank space. The box is a single form that now creates two spaces -- the inside and outside of the box. Each of these spaces is now a *part* of the original wholeness.

Once the first distinction is made, further distinctions can also be made, i.e., a sequence of boxes can be drawn, and a calculus (set of rules) can be made on how to combine the distinctions. For instance, a series of adjacent boxes can be used to count, and with the addition of a few additional simple rules, the Laws of Form can be made equivalent to number theory.(3,4) Amazingly, if it is assumed that two adjacent boxes cannot be distinguished from a single box, and a few rules, almost identical to those used in number theory, are added, the Laws of Form become equivalent to propositional logic(2). Preliminary exploration indicates that another variation of these rules can give rise to set theory.(3)

Zero versus void. Number theory includes a zero (0), and set theory includes the empty set (that which contains nothing). It might be thought that these are the same as the void, but in the Laws of Form it is important to understand that they are not the same. Zero is a concept of nothing (zero amount), and the empty set is a concept of a collection for which for which nothing has been found that fits the requirements. But a concept is a map, and a map is not the territory. Here the territory is no-thing and no-concept. A symbol for nothing is not nothing.

A basic difference between zero and void is that the void, being all-encompassing, cannot contradict anything. Thus a zero can be associated with a true or false statement, but the mystical void cannot. You might think of the void as a blank screen on which you can project pictures. If you project a symbol, like 0, it may interfere with other symbols projected on the screen. But a blank screen cannot have this effect.

As we have seen, present day science does not incorporate the idea of the void. Science studies parts and relationships, and while it has fathomed profound relations between these parts, it takes the parts as fundamental. But in the deepest analysis, parts cannot be taken as given; they must ultimately be generated by dividing wholeness. At the foundation of present day science are physical fields which are governed by abstract mathematical equations, so that the universe is explained as a web of relationships. And ultimately, it is expected that these fields will be expressed as a single unified field. But even a unified field is a thing, a distinction. By incorporating Laws of Form into the fundamental mathematics which lies at the base of science, the idea of the void can be set at its base and the implications of this basis can be explored. Laws of Form offers the vision and the means to transcend distinctions to no-thing, a nothing that is open to spirit.

* Spencer-Brown usually refers to the void as the "unmarked state," but also uses the former term.

(1) Weinless, M. (1987). The samhita of sets. *Modern Science and Vedic Science*, 1(2), 141-204.
(2) Spencer-Brown, G. *Laws of Form*.
(3) Engstrom, J.S. (1994). Natural numbers and finite sets derived from G. Spencer-Brown's *Laws of Form*. Thesis, Master of Science in Mathematics, Maharishi University of Management, Fairfield, IA.
(4) Kauffman, L.H., (1995). Arithmetic in the form. *Cybernetics and Systems*, 26, 1-57.

Acknowledgements

I thank Lou Kauffman for inviting me to write this paper...It was fitting for me to read Peirce, and to start working out some of its significance to logic and LoF and semiotics. I thank Lou Kauffman, Søren Brier, Christopher Everett, and Randhy Lockhe, both for their suggestions, which greatly improved the readability of this paper, and for their moral support. And I thank Cathy Gorini and the mathematics department of Maharishi University of Management for their training and moral support, in particular Eric Hart for computer support, and Maharishi University of Management for the benefic environment I live in.

References

Carnap, Rudolf. (1958) *Introduction to symbolic logic and its applications*, New York: Dover (27, 31).
Curry, Haskell B. (1963) *Foundations of mathematical logic*, New York: McGraw-Hill (293-294).
Encyclopedia Britannica Online , "Logic, History of; Charles Sanders Peirce" <http://www.eb.com:180/bol/topic?eu=11991&sctn=10&pm=1> [Accessed 28 February 2001]
Engstrom, Jack. (1999) "G. Spencer-Brown's *Laws of Form* as a revolutionary, unifying notation", *Semiotica* 125-1/2, (33-35, 39-40).
Engstrom, Jack. (1996a) "A notation unifying wholes and parts: G. Spencer-Brown's *Laws of Form*". See Appendix 2 of this paper.
Engstrom, Jack. (1996b) "Laws of Form and the Mystical Void". See Appendix 3 of this paper.
Peirce, Charles Sanders. (1933) *Collected Papers of Charles Sanders Peirce*, Cambridge: Harvard University Press.
Spencer-Brown, G. (1969) *Laws of Form*, London: George Allen & Unwin, Ltd. The page numbers cited here are from the New York: Julian Press 1972 edition. Republished 1994 by Cognizer, & 1997 by Bohmeier (in German).

C.S. Peirce's "First Curiosity": The World's Most Complicated Card Trick

Robin Robertson

About 1860 I cooked up a *mélange* of effects of most of the elementary principles of cyclic arithmetic; and ever since, at the end of some evening's card-play, I have occasionally exhibited it in the form of a "trick" (though there is really no trick about the phenomenon) with the uniform result of interesting and surprising all the company, albeit their mathematical powers have ranged from a bare sufficiency for an altruistic tolerance of cards up to those of some of the mightiest mathematicians of the age, who assuredly with a little reflection could have unraveled the marvel. (Collected Papers of C. S. Peirce, Vol. 4, Book III, Chapter 1).

There followed sixty pages of description and explanation of this "trick", which remained unknown by magicians until Tom Ransom rediscovered it in 1955. He in turn sent it to P. Howard Lyons, who published it as "the world's most complicated card trick," in the Nov. 1955 issue of *Ibidem*, a magazine for advanced card magicians.

It remained a curiosity, since few, if any, magicians would care to struggle through Peirce's description of cyclic arithmetic just to be able to do a relatively boring card trick. But one magician — Alex Elmsley — persevered. Elmsley had an ideal background, being both a technical systems analyst of the "bits and bites" variety, as well as one of the most brilliant creators of card magic of modern times. Elmsley isolated two principles in Peirce's trick and developed two card tricks of his own using what he called Peirce's first and second principles. I'll describe Elmsley's application of Peirce's first principle, which is presented as a humorous bet, but which could conceivably be used as an actual bar bet. The second principle, and the trick based on it, is much more complex, so we'll leave that for now.

The "con-man" — that's you — approaches someone who looks like they would be willing to make a bet; hereafter called the "sucker". You take out any 12 cards from a normal deck of playing cards. To illustrate the bet, you have the sucker pick any number between 1 and 12. Let's say he chooses 7, but any number will do. The sucker deals the cards into two face-down piles. When he reaches the 7th card in the deal, he turns it face-up, then continues dealing the rest of the cards. Finally he puts the first pile on top of the second pile. That's the procedure that is to be used throughout.

Here's the bet: if the card that falls 7th in any deal is face-up, the sucker wins $100. If it's face-down, he turns it face-up and pays you $1 the first round, $2 the next round, and so on, doubling the pay-off each time. You explain that you're giving him odds of 100 to 1, and that his chances become better each round, since there will be more and more cards face-up. The sucker is confident he'll win your $100 long before he pays you any substantial amount, so he takes the bet.

In fact, the spectator will lose 11 rounds in a row, leaving all 12 cards face-up, and will have lost $2,047 dollars in total ($1 + $2 + ... + $512 + $1024).

Why? Because this procedure creates a chain, where each card in the chain occupies every other position in the chain before it returns to its original position. Therefore, every face-down card has to appear in the 7th position (and for that matter, every position) before the original face-up card (or any other) shows up there again. Now if Peirce had thought of this betting application of his principle, perhaps he wouldn't have had to struggle with money so much in his life.

Peirce's Influence on Today's Mathematical Logic

William A. Howard[1]

Abstract: Mathematical logic has two aspects: the syntactic aspect (which, for convenience, we can regard as including the deductive aspect) and the semantic aspect. Peirce's main influence on today's mathematical logic lies in his role in the development of first-order logic. Whereas the syntactic aspect of today's mathematical logic comes to us from Frege, the semantic aspect can, in a reasonable sense, be traced back to Peirce.

Introduction

In this paper our approach is as follows. We want to know where the main ideas of today's mathematical logic came from. We are interested in the flow of ideas. Who created the ideas, or first stated them in clear form? Hence we ask: What are the main ideas of today's mathematical logic, and which (if any) can be traced back to Peirce? We will see that Peirce's influence on today's mathematical logic occurs mainly through the work of Löwenheim (1915). Although Löwenheim's work is well-known to mathematical logicians, his name may not have much significance for the non-specialist. On the other hand, the names *Frege* and *Russell* will be familiar to anyone with even a passing interest in logic or philosophy. The viewpoint of most accounts of the history of logic is that today's logic originated with Frege and comes to us via Russell, perhaps with a contribution from Peano. This ignores the fact that, in the late nineteenth and early twentieth century, there were two lines of development of mathematical logic: Frege-Russell and Peirce-Schröder-Löwenheim. To understand Peirce's influence on today's logic, it is necessary to have some understanding of the ideas involved in these two lines of development. Hence in sec. 1-4 we look at some of the basic ideas underlying mathematical logic. Then, in sec. 5-8, we look at the two lines of development, with emphasis on the Peirce-Schröder-Löwenheim line.

The centerpiece of today's mathematical logic is a system of logic called *first-order logic*. Peirce's main influence on today's mathematical logic lies precisely in his influence on the development of first-order logic. We devote sections 9-11 to a discussion of first-order logic and the nature of Peirce's contributions to it. The main conclusion of this paper is: The semantic aspect of today's first-order logic can be traced back to Peirce, in a reasonable sense. The qualifier "in a reasonable sense" has been included because today's first-order

---

[1] University of Illinois at Chicago, Department of Mscs (M/C 249), 851 South Morgan Street, Chicago, Illinois 60607-7045, USA. *E-mail:* wahow@uic.edu

logic may have more than one source (i.e., a source other than Löwenheim's 1915 paper). This is discussed in sec. 11-12.

1. What is logic?

Peirce, in his 1896 review of a work by Schröder defined logic as follows.

> "Logic may be defined as the science of the laws of the stable establishment of beliefs." (Hartshorne and Weiss 1933, p. 271)

This is an unusually comprehensive view of logic. It goes beyond what most people would mean by 'logic'. Webster's New Universal Dictionary defines logic as follows.

> "[Logic is] the science which investigates the principles governing correct or reliable inference."

This is, for our purposes, too narrow because—in the area of mathematical logic, at least—logic has not only a deductive aspect but also a semantic aspect. (See next two sections.)

2. The two meanings of the phrase 'mathematical logic'

The primary meaning of the phrase *mathematical logic* is: logic is to be formulated and studied in a mathematical way. As part of this process, logic is developed by use of a symbolic language. This is what Peirce would call *formal logic*. The secondary meaning is: the logic to be studied consists of the logic upon which mathematical proofs are based. In principle, the methods used in studying logic according to the second meaning need not themselves be mathematical; but, starting with Frege and Peirce, this area has in fact been dominated by mathematical methods. In the work of Boole and Peirce, the first meaning predominates. This tradition goes back to Leibniz: logical reasoning is to be carried out by calculating in a symbolic system. In particular, an argument between philosophers would be resolved by formulating the issue symbolically, then calculating the answer.

In the work of Frege, the second meaning of the phrase 'mathematical logic' predominates. Frege's goal was to provide a foundation of mathematics based on a theory of concepts. To develop this theory, he introduced a special symbolic language and supplied it with rules of deduction. (For convenience, we can take one of the rules to be: assert a "logical axiom".) One reason he needed rules of deduction is the following. Concepts are a part of logic. Hence, in using logic to carry out proofs in a theory of concepts, there is a danger of circularity. This circularity is avoided if the proofs can be carried out in a purely mechanical fashion, each step being made by applying one of the rules.

Whereas for Frege and Russell, logic was to provide a foundation of mathematics, Peirce did not see a need for a foundation of mathematics (and, in fact, saw mathematics as prior to logic). To put this succinctly, one might say: whereas Frege wanted to turn mathematics into logic, Peirce wanted to turn logic into mathematics (i.e., make logic look more like mathematics).

3. Syntax vs. semantics

In the deductive aspect of logic, there are rules for deducing sentences B from sentences A. More generally, sentences B are deduced from a finite set of premises C, D, ...; but we can take A to be the conjunction of these premises. If B is deduced from A, the argument proceeds as follows: "If you grant me that A is true, then you must grant me the truth of B." This goes back to the ancient Greeks (in particular, Plato and Aristotle). Syntax deals with the structure of the expressions of a language. A proof is a (finite) sequence of expressions satisfying appropriate conditions. Since a sequence of expressions can itself be regarded as an expression, the deductive aspect of logic can be included in the syntactic aspect.

Semantics is the study of the relation between expressions and the objects they talk about. Thus the notion of truth (of a sentence) is a semantic notion. As already indicated, there is a connection between the deductive and semantic aspects of logic; namely, the justification of a rule of deduction is that it preserves truth. In fact, one might regard the rules of deduction as a codification of the relation R(A, B) which says that whenever A is true, B is also true.

4. Some connections with second-order cybernetics

In talking about rules of deduction, the Greeks took the step from the *use* of the rules to the *study* of the rules. Thus the rules become the new subject matter. For example, the method of proof by contradiction depends on this *step of reflection* (which is why students find it hard today). In performing a step of reflection, we draw back and look at what we have been doing. We move from *inside* the system to the *outside*. This is a fundamental process of human thought. In examining *mathematical* thought, we see this process in a clear-cut form.

If we are able to work within a system of thought while simultaneously viewing it from outside, this would be an example of second-order cybernetics: the observer is part of the system.

Corresponding to the process of thinking about thoughts, we have language talking about language. This always happens in mathematical logic. Language #1 talks about (for example) mathematical objects; and language #2 talks about both language #1 and the mathematical objects. Why not use just one language; in other words, take language #1 and language #2 to be the same? Then we would have a language that talks about itself. The difficulty is that such a language cannot talk in a coherent way about the truth of its own sentences. In more detail, if L is a language with a reasonable ability to talk about natural numbers, then L cannot have a symbol T such that, for every sentence A of L, T(A) says that A is true. This follows from the fact that, for any such language L, we can produce a sentence which can be neither true nor false (Tarski's parallel to Gödel's incompleteness theorem). The situation described in the present paragaph is a semantic counterpart of what Varela calls *Gödel's loop* (see page 315 of Varela, 1984).

5. From Boole and De Morgan to Peirce

Boole's first breakthrough was his discovery of how to develop an algebra of classes. Leibniz had tried to do this but was not very successful, perhaps because of a tendency to think in terms of predicates (properties) rather than classes. Boole's second breakthrough was his discovery that his algebra of classes could be reinterpreted as an algebra of propositions, or more precisely, an algebra of truth functions. Since truth and falsehood are represented in a convenient way by the numbers 1 and 0 (so conjunction of propositions corresponds to multiplication of their truth values), Boole's theory of propositions had a particularly mathematical appearance. One can see that Boole's analysis of propositions was fundamental: he analyzed the behavior of truth in relation to the operations of conjunction, disjunction and negation.

Boole's work appeared first in 1847, and then, in fuller form, in 1854. At the same time, De Morgan was concerned with the fact that an adequate logic must treat not only classes but also relations. He introduced the idea of the *relative product* of binary relations; from relations A and B, we form the relation C such that C(i, j) means: there exists k such that A(i, k) and B(k, j) both hold. For example, from the relations *father of* and *parent of* we get the relation *grandfather of*. Peirce's first big accomplishment in mathematical logic was his discovery of how to combine De Morgan's notion of relative product with Boole's algebra of classes, thus producing an algebra of relations (Peirce, 1870).

6. Peirce's introduction of the quantifiers

Peirce's algebra of relations, as a system of general logic, has the following severe limitation. One cannot, in general, take the negation of a universal statement. In other words, the ability of the algebra of relations to handle existential statements is severely limited. What was needed was a general method of handling the ideas of *for all* and *there exists*. His student, O. H. Mitchell, provided a solution, and this inspired Peirce to provide his own solution. He developed this in his paper of 1885. Peirce's solution is based on the idea that if we are considering a property A(i) of finitely many individuals i = 1, 2, 3, ..., n, then to say "A(i), for all i" is equivalent to saying "A(1) and A(2) and A(3) and ...

and A(n)". Since, in the Boolean tradition, conjunction is represented by multiplication, the latter can be written as $\Pi_i A(i)$; but Peirce used the notation $\Pi_i \alpha_i$ since he was thinking of propositions a_i indexed by a variable i. Guided by this, he switched his viewpoint and, for an arbitrary domain of individuals (finite or infinite), he took $\Pi_i a_i$ to mean, simply: the proposition a_i holds for every individual i. Similarly, he introduced the quantifier Σ_i to represent existential quantification. Since these ideas go in a direction different from Mitchell's ideas, it is correct to say that Peirce introduced the quantifiers. This does not shortchange Mitchell: he had a form of quantification but he did not have the quantifiers.

Thus, in his paper of 1885, Peirce introduced a symbolic language based on individual variables, predicate symbols, propositional connectives and quantifiers. On the deductive side, he provided various rules of calculation by means of which new expressions can be deduced from given expressions. All this is done in his section III, which is devoted to what today would be called first-order logic. In section IV he extends this to second-order logic; namely, there are second-order predicates (i.e., predicates applied to first-order predicates), and quantifiers whose variables range over first-order predicates. The paper of 1885 represents Peirce's second big accomplishment in mathematical logic.

Since Frege had already introduced universal quantification in his monograph of 1879, the question arises as to whether Peirce's work was independent of Frege. As far as I can see, the answer is "Yes". For a discussion of this and other issues concerning Peirce's introduction of the quantifiers, see Brady (1990).

7. From Peirce to Schröder to Löwenheim

Schröder, in Vol. III (1895) of his *Vorlesungen über die Algebra der Logik*, adopted Peirce's systems of first- and second-order logic. He replaced Peirce's computational system of deduction by various rules (or axioms) resembling those of today's logic. But actually, Schröder's work was stronger on the semantic side than it was on the deductive side. Löwenheim's 1915 paper is based on Schröder's Vol. III. A brief examination of Löwenheim's paper and Schröder's Vol. III shows that the framework of logic within which Löwenheim worked was supplied by Peirce's 1885 paper (via Schröder's Vol. III). Löwenheim's paper of 1915 played a central role in the development of today's mathematical logic.

8. The Frege-Russell line of development

Russell developed a symbolic language that combined the best features of the symbolic languages of Frege and Peano; namely, he combined the precision and conceptual clarity of Frege's symbolic language with the readability of Peano's symbolism. Also, Whitehead and Russell adopted a modified version of Frege's rules of deduction. The modifications are inessential. (Schröder may have had some influence; at least, some of the logical axioms and derived rules of inference in the work of Whitehead and Russell are remarkably similar to some of the formulas in Schröder's Vol. III.) This language, and rules of deduction, provide the basis of the massive three-volume work: Whitehead and Russell, *Principia Mathematica,* 1910-1913.

9. Today's mathematical logic

The centerpiece of today's mathematical logic is *first-order logic*. In a *first-order language*, the atomic expressions are of the form A(r), B(r, s), C(r, s, t), etc., where r, s, t, ... are individual variables or constants, and the letters A, B, C, ..., called predicate letters, are intended to denote properties or relations. (For the present purpose, we do not need to consider languages with function signs.) The formulas of the language are built up from the atomic formulas by use of the propositional connectives and the quantifiers

$$\forall, \exists$$

applied to individual variables. A quantified variable is said to be *bound*, otherwise *free*. A formula is said to be *closed* if it has no free variables. The language is said to be *first-order* because there are no predicates applied to predicates (i.e., properties are not treated as individuals), and there is no quantification over predicates).

Rules of deduction
Although Frege and Whitehead-Russell did not consider first-order logic separately, their rules carry over. In this essay, the phrase 'rules of deduction' is to be understood in the light of the fact that in most presentations of logic, many of the rules of deduction are encoded as logical axioms, and these are unwound by means of *modus ponens*:

from A and

$$A \Rightarrow B$$

infer B. For example, the inference

$$\text{From A infer } A \vee C$$

follows from the logical axiom

$$A \Rightarrow A \vee C$$

and modus ponens.

10. Semantics of first-order logic

A closed formula of a first-order language does not, by itself, have a meaning. It acquires a meaning when the language is provided with an *interpretation*. An interpretation of a first-order language is obtained by choosing a domain D of individuals and by assigning specific relations on D to the predicate letters (and assigning individuals to the constants, if the language contains constants). Consider, for example, the formula

$$(10.1) \qquad \forall x \forall y (B(x,y) \Rightarrow B(y,x)).$$

Interpretation 1
D is the set of all positive integers and B(x, y) is taken to mean x < y. Under this interpretation, (10.1) is false.

Interpretation 2
D is the set of all real numbers and B(x, y) is taken to mean x + y = 1. Under this interpretation, (10.1) is true.

A closed formula of a first-order language is said to be *logically valid* if it is true under all interpretations. Thus the formula

$$(10.2) \qquad \forall x \exists y B(x,y) \Rightarrow \exists y \forall x B(x,y)$$

is not logically valid because it is false under interpretation 1, above. On the other hand, the formula

$$(10.3) \qquad \exists y \forall x B(x,y) \Rightarrow \forall x \exists y B(x,y)$$

is logically valid. To explain this in Peirce's manner, let D be a group of people and suppose B(x, y) says "x loves y". Then the antecedent of (10.3) says that there is a person *p* whom everybody loves. From this, the consequent of (10.3) follows because, for each person x in the group, there is someone whom x loves; namely, *p*. But the same reasoning applies to any domain of individuals, nomatter what B(x, y) means. The example (10.3) appears in Peirce's 1885 paper in the form

$$(10.4) \qquad \Sigma_k \Pi_i b_{ik} \prec \Pi_i \Sigma_k b_{ik}$$

(except for different letters).

Logical consequence
A closed formula G is said to be a logical consequence of a set of closed formulas P, Q, ... if every interpretation that makes P, Q, ... true also makes G true.

11. The origins of today's first-order logic

Today's first-order logic comes to us from two sources. The first source is the Peirce-Schröder-Löwenheim line of development (see section 7, above). We have already mentioned (sec. 6) that sec. III of Peirce's 1885 paper is devoted to first-order logic. Indeed, Peirce's sec. III is entitled *First-intentional Logic of Relatives*. 'Relative' was Peirce's word for *relation*. Moreover, for Peirce a relation was represented by a set of n-tuples of elements taken from an arbitrary set of individuals. This is the extensional view of relations. Thus there is no doubt that Peirce had today's semantics of first-order logic. In the following section, he developed second-order logic, which he called *second-intentional logic*. He took this terminology from scholastic philosophy, in which he was a world-class authority. *Second-intention* means: thoughts are considered as objects (a relation is a thought; i.e., a mental object). Thus Peirce had a clear philosophical reason for separating first- and second-order logic. This separation was preserved by Schröder and was crucial to Löwenheim's 1915 paper. The ideas of Löwenheim's paper were clarified and further developed by Skolem in papers of 1920 and 1923.

The second source of today's first-order logic consists of Hilbert's 1917-1918 lectures. But in Hilbert's development, the emphasis is on the syntactic side; the treatment of the semantic side is weak. The first systematic presentation of today's first-order logic occurs in the book of Hilbert and Ackermann (1928); so we need to ask: In the book of Hilbert and Ackermann, where did the ideas concerning the semantics of first-order logic come from? My impression is that the development of first-order logic in the period 1917–1928 was strongly influenced by Löwenheim's 1915 paper. But this is a question that needs to be investigated.

12. Tracing back the ideas of first-order logic

Van Heijenoort (1976) pictures the two lines of development, Frege-Russell and Peirce-Schröder-Löwenheim, as constituting two streams: the syntactic stream and the semantic stream, respectively. He points out that in 1920 the two streams begin to "mix their waters". In looking for the origins of first-order logic, this mixing of the waters must be taken into account. Nonetheless, we can say the following.

Today's notation is essentially that of Whitehead-Russell (see sec. 8). Actually, in Whitehead-Russell, some of Frege's precision and conceptual clarity was lost. It was restored by Hilbert (1917-1918)—probably not by going back to Frege but, rather, simply by seeing what was needed. Also, as is well-known, Frege and Whitehead-Russell did not single out first-order logic as an autonomous system; this was done by Hilbert (but also Peirce treated first-order logic as an autonomous system in his 1885 paper).

Under the heading of *syntax* (i.e., formalism) of a system of mathematical logic, we include not only the structure of the language but also the rules of deduction (see sec. 3). Thus:

> (12.1)   The syntactic aspect of today's first-order logic can be traced back to Frege.

We can also say:

(12.2)   The semantic aspect of today's first-order logic can be traced back to Peirce, in a reasonable sense.

This is because of Peirce's influence on the work of Löwenheim and Skolem, and the importance of the work of Löwenheim and Skolem for today's mathematical logic. The qualifier "in a reasonable sense" has been included because the role of the book of Hilbert and Ackermann (1928) has to be taken into account. If we want to give Peirce as much credit as possible, the worst case would be that the work of Löwenheim and Skolem had no influence on the work leading to the presentation of first-order logic in the book of Hilbert and Ackermann. Actually, I suspect that Löwenheim and Skolem had an important influence on the work just mentioned, in which case the qualifier in (12.2) could be dropped. This needs to be investigated.

13. Peirce, Löwenheim, and model theory

Although the semantics of today's first-order logic has more than one source, the work of Löwenheim (and hence Peirce) is an important source. This is because the work of Löwenheim is crucial to a branch of today's logic known as model theory. By a *model* of one or more closed formulas of a first-order language is meant an interpretation which makes these formulas simultaneously true (see sec. 10). In model theory, the objects of study are models rather than proofs or concepts. The model-theoretic development of logic is based on what van Heijenoort (1986) calls "set-theoretic semantics". The required set theory is regarded as given. Thus the model theorist is not concerned with the foundations of mathematics. Indeed, the viewpoint of model theory toward the relation between mathematics and logic is much the same as Peirce's (see sec. 2).

14. A tardy discovery

During the years leading up to the 1885 paper, Peirce struggled to clarify the idea of "logical necessity". What he needed was the notion of *logical consequence* (see the end of sec. 10, above). The notions of logical validity and logical consequence are implicit in the 1885 paper, but Peirce was still struggling with these ideas and was unable to make them explicit. Finally in his 1896 review of one of Schröder's volumes he was able to define the notion of logical consequence (Hartshorne and Weiss 1933, p. 278; for a discussion of this passage in Peirce's review, see pp. 54-55 of Iliff, 1992). This was an excellent accomplishment; but nobody noticed it, so I have not included it in the main part of this essay, which is devoted to an account of Peirce's influence on today's mathematical logic.

References

Brady, G. 1990, *Peirce's Introduction of the Quantifiers*. M.A. Thesis, Department of Philosophy, University of Chicago.

Hartshorne and Weiss 1933, *Collected Papers of Charles Sanders Peirce*, Vol. 3 (Harvard University Press).

Hilbert, D. 1917, *Prinzipien der Mathematik und Logik*, Lecture notes of a course given at Göttingen, 1917-1918 (Math. Institut, Göttingen).

Hilbert and Ackermann 1928, *Grundzüge der Theoretischen Logik,* (Berlin: Springer).

Iliff, A. J. 1992, *Charles S. Peirce's Contributions to Mathematical Logic and Philosophy*. D.A. Thesis, University of Illinois at Chicago.

Löwenheim, L. 1915, "On Possibilities in the Calculus of Relatives," in *From Frege to Gödel: A Source Book in Mathematical Logic, 1879-1931,* ed. Jean van Heijenoort (Harvard University Press, 1967), pp. 228-251.

Peirce, C. 1870, "Description of a Notation for the Logic of Relatives, Resulting from an Amplification of the Conceptions of Boole's Calculus of Logic," in Vol. 2 (1984) of *Writings of Charles S. Peirce: A Chronological Edition,* ed. Max Fisch (Indiana University Press).

——. 1885, "On the Algebra of Logic," *The American Journal of Mathematics*, **7**, 180-202.

Schröder, E. 1995, *Vorlesungen über die Algebra der Logik*, Vol. III (Leipzig: Teubner).

van Heijenoort, J. 1976, "Set-Theoretic Semantics," in: van Heijenoort, *Selected Essays* (Naples: Bibliopolis, 1985).

Varela, F. 1984, "The Creative Circle: Sketches on the Natural History of Circularity," in: *The Invented Reality*, ed. Paul Watzlawick (W. W. Norton, New York, 1984).

The Mathematics of Charles Sanders Peirce

Louis H. Kauffman[1]

I. Introduction

This essay explores the Mathematics of Charles Sanders Peirce. We concentrate on his notational approaches to basic logic and his general ideas about Sign, Symbol and diagrammatic thought.

In the course of this paper we discuss two notations of Peirce, one of Nicod and one of Spencer-Brown. Needless to say, a notation connotes an entire language and these contexts are elaborated herein. The first Peirce notation is the portmanteau (see below) Sign of illation. The second Peirce notation is the form of implication in the existential graphs (see below). The Nicod notation is a portmanteau of the Sheffer stroke and an (overbar) negation sign. The Spencer-Brown notation is in line with the Peirce Sign of illation. It remained for Spencer-Brown (some fifty years after Peirce and Nicod) to see the relevance of an arithmetic of forms underlying his notation and thus putting the final touch on a development that, from a broad perspective, looks like the world mind doing its best to remember the significant patterns that join logic, speech and mathematics. The movement downward to the Form ("we take the form of distinction for the form."[9, Chapter 1, page 1]) through the joining together of words into archetypal portmanteau Signs can be no accident in this process of return to the beginning.

We study a system of logic devised by Peirce based on a single Sign for inference that he calls his 'Sign of illation". We then turn to Peirce's Existential Graphs. The Existential Graphs lead to a remarkable connection between the very first steps in Logic and mirror plane symmetries of a "Logical Garnet" [30] in three dimensional space. Peirce's ideas about these graphs are related to his ideas about infinity and infinitesimals, and with his more general philosophy that regards a human being as a Sign. It is the intent of this paper to bring forth these themes in both their generality and their particularity.

It is amazing that three dimensional geometry is closely allied to the first few distinctions of Logic. It is my intent in this paper to make that aspect crystal clear.

---

[1] Department of Mathematics, University of Illinois, Chicago. Email: kauffman@uic.edu. I would like to take this opportunity to thank Diane Slaviero, David Solzman, Jim Flagg, G. Spencer-Brown, Annetta Pedretti and Kyoko Inoue for many conversations real and imaginary related to this paper. I wish to dedicate this paper to the memory of Milton Singer and to our many meetings in the Piccolo Mondo Restaurant in Hyde Park, Chicago in the 1990's.

We also clarify the relationships of Peirce's Mathematics in other ways that are described below. These clarifications are the specific content of the paper. Their purpose is to shed light on the beautiful philosophical generality of Peirce's work, and to encourage the reader to look at this in his or her own context and in the contexts of semiotics and second order cybernetics.

We begin with a detective story about the design of a Sign for inference that Peirce calls the "Sign of illation." Peirce designed this Sign as a "portmanteau", a combination of two Signs with two meanings! The plot thickens as we find that the very same design idea was independently taken up by two other authors (Nicod and Spencer-Brown). This is treated in Section 2 of this paper. The Sign of illation is a convenient Sign for inference. A very similar combination Sign was independently devised by Nicod, using the Sheffer stroke and overbar negation. I illustrate these notations and how they are related. The way that both Peirce and Nicod arrive at a portmanteau symbol is part of a movement that goes beyond either of them to unify logic and logical notation into a simpler structure.

The double meaning of the portmanteau is a precursor to the interlock of syntax and semantics that led to Gödel's work on the incompleteness of formal systems. See Section 11 for a discussion of this theme.

Peirce also had another approach to basic logic. This is his theory of Existential Graphs. Peirce's Existential Graphs are an economical way to write first order logic in diagrams on a plane, by using a combination of alphabetical symbols and circles and ovals. Existential graphs grow from these beginnings and become a well-formed two dimensional algebra. I make the following observation: *There is a natural combinatorial "arithmetic" of circles and ovals that underlies the Peircian Existential Graphs.* The arithmetic of circles is a formal system that is interpreted in terms of itself. It is a calculus about the properties of the distinction made by any circle or oval in the plane, and by abduction it is about the properties of any distinction. This circle arithmetic in relation to existential graphs is discussed in Section 6 where we show that it is isomorphic with the Calculus of Indications of G. Spencer-Brown. Spencer-Brown's work can be seen as part of a continuous progression that began with Peirce's Existential Graphs. In essence what Spencer-Brown adds to the existential graphs is the use of the unmarked state. That is, he allows the use of empty space in place of a complex of Signs. This makes a profound difference and reveals a beautiful and simple calculus of indications underlying the existential graphs. Indeed Spencer-Brown's true contribution is that he added Nothing to the Peirce theory!

In Section 7 we discuss how the "Logical Garnet" of Shea Zellweger fits into this picture. Zellweger discovered that the sixteen binary connectives studied by Peirce fit naturally on the vertices of a rhombic dodecahedron (plus a new central vertex) in such a way that symmetries of these connectives correspond to mirror symmetries of this polyhedron in three dimensional space. The Logical Garnet fits perfectly into the context of the existential graphs.

In Section 8 we discuss Peirce's ideas about continuity and infinitesimals, and relate this to extensions of existential graphs to infinite graphs. In Section 9 we quote a famous passage in Peirce about a "Sign of itself" and discuss this passage in terms of topology and self-reference. This passage and the remarks on infinitesimals show how Peirce's thought reaches far beyond the specific formalisms that he produced, and that his intuition was right on target with respect to much of the subsequent (and future!) development of mathematics and logic. Section 10 continues the discussion of Section 9 in the context of second order cybernetics. Section 11 is an epilogue and a reflection on the theme of the portmanteau word.

II. The Sign of Illation

Peirce wrote a remarkable essay [1] on the Boolean mathematics of a Sign that combines the properties of addition and negation. It is a portmanteau Sign in the sense of Lewis Carroll (See below and Section III). We do not have the capabilities to typeset the Peirce Sign of illation, but see right for a rendition of it.

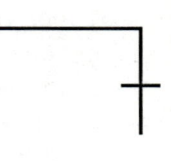

Instead, I shall use the following version in this text: [a]b. When you see [a]b in the text you are to imagine that a horizontal bar has been placed over the top of the letter "a", and a vertical bar, crossed with a horizontal bar (very like a plus sign) has been placed just to the right of the "a" in such a way that the vertical bar and the horizontal overbar share a corner. In this way [a]b forms the Peirce Sign of illation, and we see that this Sign has been created by fusing a horizontal bar with a plus sign. The horizontal bar can be interpreted as negating the Sign beneath it.

Peirce went on to write an essay on the formal properties of his *Sign of illation* and how it could be used in symbolic logic. Here is the dictionary definition of the word illation. Note that we have taken the "Sign of Illation" as the title of this section of the paper.

il . la`tion. [L. *illation*, fr. *illatus*, used as past participle of *inferre*, to carry or bring in] Inference from premises of reasons; hence that which is inferred or deduced. [2]

$$A \mathbin{\top} B = \overline{A} + B$$

$$\overline{A} = \text{not } A$$

The Sign of illation enables a number of notational conveniences, not the least of which is that the implication "A implies B" usually written as " A -> B" is expressed as "[A]B " using the Sign of illation.

The Sign of illation is a *portmanteau Sign* in the sense of Lewis Caroll [3], [4] who created that concept in his poem "Jabberwocky" where one encounters a bestiary of words like "slithy" – a combination of lithe and slimy. A portmanteau is literally a coat and hat rack (also a suitcase), an object designed to hold a multiplicity of objects. Just so, a portmanteau word is a holder of two or more words, each justly truncated to fit with the truncate of the other. A modern version to contemplate is the word "smog" a combination of "smoke" and "fog".

It is this fitting together of the two words that is so characteristic of the portmanteau. It recalls the amazing doublings of nature that would use a mouth for eating and speaking, a throat for breathing and drinking, and the amazing multiple use of the DNA at biology's core. In Peirce's case of his portmanteau Sign of illation there is no truncation, but rather a perfect fit at the corner of the horizontal overbar as sign of negation, and the vertical plus sign as sign of "logical or". The two fit into one Sign that can then hold neatly yet another meaning as a Sign of implication.

It is the meaning of the Sign of illation as implication that Peirce takes as primary. In his essay [1], he begins with this interpretation, deduces many properties of the Sign from this interpretation, and only in the last paragraph does he reveal that his Sign can be taken apart into a plus sign and overbar (interpreted as a negation). In the beginning he writes [1]

> This symbol must signify the relation of antecedent to consequent. In the form I would propose for it, it takes the shape of a cross placed between antecedent and consequent with a sort of streamer extending over the former. Thus, "if a then b" would be written [a]b.

$$\overline{A} + B$$

> From a, it follows that if b then c", would be written [a][b]c.

$$\overline{A} + \overline{B} + C$$

> "From 'if a then b.' follows c," would be written [[a]b]c.

$$\overline{\overline{A} + B} + C$$

> To say that a is false, is the same as to say that from a as an antecedent follows any consequent that we like. This is naturally shown by leaving a blank space for the consequent, which may be filled in at pleasure. That is, we may write "a is false" as [a],

$$\overline{A} +$$

> implying that from a, every consequence may be drawn without passing from a true antecedent to a false consequent, since a is not true." [We are still quoting from [1].]

At this point, Peirce has partially let the cat out of the bag by noting that with the use of a blank space for a variable, his Sign can express negation.

He then introduces signs 0, 1, to stand for falsity (absence) and truth (presence) respectively. The symbol 1 is taken to stand in for the expression [a]a for any a (as 'a implies a" is true for any a).

$$\overline{A} + A$$

One of the charming features of the essay is that he deduces many formal properties of this symbolism wholly conceptually, based on this interpretation of the Sign of illation as implication. In fact, he remarks [1] " Here then we have a written language for relations of dependence. We have only to bear in mind the meaning of the symbol

$$\overline{A\mid} B$$

(not by translating it into if and then, but by associating it directly with the conception of the relation it signifies), in order to reason as well in this language as in the vernacular, - and indeed much better."

At the end of the essay [1], he writes

> We now have a complete algebra for qualitative reasoning concerning individuals. But it is not yet a very commodious calculus. To render it so, we introduce certain abbreviations which make it identical with the algebra of Boole ... Namely, we first separate the streamer of the Sign of illation from the cross, and in place of [a]b write ~a + b.

$$\overline{A\mid} B = \sim A + B$$

Second,
whenever the Sign of illation is followed by a blank we omit the cross, and thus in place of [a], write ~a.

$$\overline{A\mid} = \sim A$$

Third,
as the sign of the simultaneous truth of a and b, instead of writing [[a][b]], we simply write ab." [1]

$$\overline{\overline{A\mid}\ \overline{B\mid}\mid} = AB$$

In the end, it is important that the portmanteau Sign can be decomposed back into its component parts, for this allows the translation between Peirce's thought and the Boolean algebra. It is these issues of translation, from one formalism to another and from meaning in natural language to meaning in the formalism, that he holds with great sensitivity.

III. Nicod and the Sheffer Stroke

A remarkable paper by Nicod [5], creates a portmanteau Sign for implication almost identical to that of Peirce. Nicod wrote in the context of the Sheffer stroke a|b

$$A|B = \text{not}(A \text{ and } B)$$

that represents "not both a and b". Nicod noticed that he could create a Sign for implication by putting a negation bar over one of the variables of the stroke. Thus "not both a and not b" is logically equivalent to "a implies b".

$$A|\overline{B} = \text{not}(A \text{ and not } B)$$
$$= A \text{ implies } B$$

Thus for Nicod

$$A\mid\overline{B}$$

stood for "A implies B" and became a "convenient sign for implication". The decomposition is nearly the same as in the Peirce Sign of illation; the meanings have shifted, and the Cheshire cat is smiling in the background.

What we are witnessing here is the peculiar relationship between spoken and written language (ordinary language) and symbolic logic.

We think naively that there is not much more to reasoning than the simplest properties of inference. That all you need to know to reason is that if A implies B and B implies C, then A implies C. And so one might start down the road to symbolic logic by a convenient sign for implication. This is just what Peirce did in his essay [1] that we discussed in the previous section. And yet when you think about the matter of when an implication is true and when it is false and how it interfaces with the meanings of "and" and "or" a complexity arises. This complexity expands to the curious intricacy of first order symbolic logic, and then it seems like a breathe of fresh air to find the symbols of the logical system combining (almost of their own accord) to assemble a sign for illation. It is clear that this experience occurred to both Peirce and to Nicod quite separately, and we shall see that they were not alone!

I believe that it is quite significant to see the sign of implication as a complex sign composed of other logical signs. This places it in its proper context. Implication itself is not simple, yet something simple underlies it. Inference is a portmanteau, a gluing of separate meanings into a coherent whole.

IV. Pivot and Portmanteau

Along with the concept of a portmanteau word or symbol there is a notion that I like to call a "pivot duality". A portmanteau word is a combination of separate meanings such that their signs can be fitted together. In a pivot duality a word or symbol can be interpreted in more than one way, and this multiplicity of interpretation gives rise to a pivot, or translation, between the different contexts of these interpretations.

Pivot duality is the essence of multiplicity of interpretation, while pormanteau is the exemplar of the condensation of a multiplicity of meanings into a single sign. A portmanteau always has an associated pivot, but a pivot need not be a portmanteau. By bringing forth the pivot, we can expand the context of the consideration of the multiplicity of meanings associated with Signs. This has direct bearing on the understanding of the use of Signs in Peirce and in language as a whole.

A good example of pivot duality is the simple Feynman diagram

$$\mathord{>}\mathord{-\!\!\!-\!\!\!-}\mathord{<}$$

that can be interpreted (with time's arrow going up the page) as two particles interacting by the exchange of a photon (the horizontal line), and can also be interpreted (with time's arrow going from left to right) as a particle and an antiparticle annihilating to produce a photon that then momentarily decays into a new particle pair. Here we have two completely different (yet related) interpretations of the same bit of formalism. The formalism seems to point to a deeper reality, beyond the particular way that the physicist observer decomposes process into space and time.

There is an affinity between the portmanteau symbol and a pivot. The portmanteau is a single word that holds two meanings. The pivot is a word or symbol or text that is subject to a multiplicity of interpretations. We make them both because these makings are the essence of the condensation of meaning into Signs and the use of Signs in the expansion of meaning. If the meaning of a Sign is its use, then the meaning of the Sign is not one but many, according to its uses, and yet one according to the unity that these uses find in the Sign itself (as a complex of Signs fully embedded in language). A hammer makes a good example, being one tool and yet being capable of both the impulsive insertion of the nail and the levering extraction of the nail. Two meanings pivot over the hammer. The combination of claw and hammerhead makes the tool itself a portmanteau of these two actions.

The reason, I believe, that portmanteau and pivot are so important to find in looking at formal systems, and in particular symbolic logic, is that the very attempt to make formal languages is fraught with the desire that each term shall have a single well assigned meaning. It cannot be! The single well-assigned meaning is against the nature of language itself. All the formal system can actually do is choose a line of development that calls some entities elementary (they are not) and builds other entities from them. Eventually meanings and full relationships to ordinary language emerge. The pattern of pivot and portmanteau is the clue to this robust nature of the formal language in relation to human thought and to the human as a Sign for itself.

The grin of the Cheshire cat is the quintessential pivot, yet it is not a portmanteau. To quote Martin Gardner in his comment on Alice's encounter with the Cheshire Cat [4, p. 91],

> The phrase 'grin without a cat' is not a bad description of pure mathematics. Although mathematical theorems can often be usefully applied to the structure of the external world, the theorems themselves are abstractions that belong in another realm 'remote from human passions,' as Bertrand Russell once put it in a memorable passage, 'remote even from the pitiful facts of Nature...an ordered cosmos where pure thought can dwell as in its natural home.'

In mathematics the grin without the cat is often obtained through a process of distillation. The structure is traversed again and again and each time the inessential is thrown away. At last only a small and potent pattern remains. This is the grin of the cat. That grin is a pattern that fits into many contexts, a key to many doors. It is this multiplicity of uses for a single symbolic form that makes mathematics useful. It is the search for such distillation of pattern that is the essence of mathematical thought.

V. Peirce's Existential Graphs

We now turn to a development of Peirce for logic that is closely related to the Sign of illation. These are his existential graphs [6], [7], [8]. In this development, Peirce takes the operations "and", "not", and a space in which they are represented as fundamental. He develops logic from that ground. It is important to see this development both for the structure of basic logic and for the view that it gives of Peirce's thought as he examines the same(!) subject from a different angle and finds that it is a different subject.

The first stage of the Existential Graphs are called the "alpha" graphs. These alpha graphs are concerned with the logic of implication, and we shall concentrate on their structure.

Here is a quick description of the context for the Existential Graphs.

We are given a plane on which to make inscriptions. If we place a graph or symbol on the plane, then the proposition corresponding to this symbol is asserted. If we place two disjoint complexes of symbols on the plane, call them A and B, then we are asserting the conjunction "A and B".

$$A \ B$$

A circle (or simple closed curve) drawn around a symbol changes the assertion to the negative. Thus a circle around A asserts the negation of A.

$$(A) = \text{not } A$$

For ease of notation, we shall make an algebraic version of the existential graphs where AB denotes "A and B" and (A) denotes a circle around A. Hence (A) denotes "not A".

Since "A implies B" is logically equivalent to "not (A and not B)" we see that "A implies B" has an existential graph consisting of a big circle that contains both A and a circle around B. Algebraically this is (A(B)) for "A implies B".

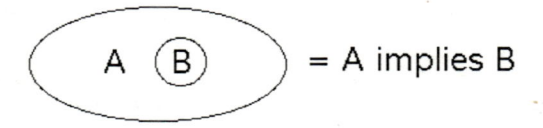

$$(A\ (B)) = A \text{ implies } B$$

Peirce worked out a number of rules for manipulating these graphs by different patterns of substitution and replacement. With such rules in place, the graphs become an arena for analyzing basic arguments and tautologies in logic. Note that the idea behind the existential graphs and the Sign of illation is essentially the same, although the underlying model for implication is "(not a) or b" in the case of the Sign of illation.

It is interesting to see how these graphs work. To this end let us set up some rules for manipulating the graphs. We must remark that in the discussion that follows I will consider only transformations that preserve the full logical structure of the Existential Graphs. Peirce considers transformations on the graphs that preserve truth in one direction (e.g. see [7]). That is he allows two graphs X and Y to be transformed one to another so long as whenever X is true then Y is true. Here we require that X is true if and only if Y is true. Since this is the usual notion of equality of logical expressions, I believe its use will introduce a measure of clarity in the study of the existential graphs.

1. Since "P implies P" is always true, we see that any graph having the pattern (P(P)) will have the truth value true.

$$(P(P)) = T$$

2. Two circles around any P has the same truth value as P: ((P)) = P.

$$((P)) = P$$

3. PQ = QP because they are both true exactly when A is true and B is true.

With the help of these patterns, we can see how various tautologies arise. For example, "A implies B implies [not B implies not A] " is transcribed to:

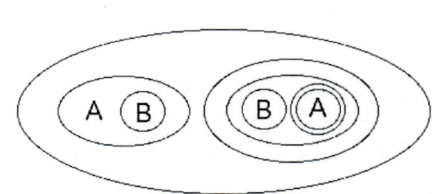

This reduces to:

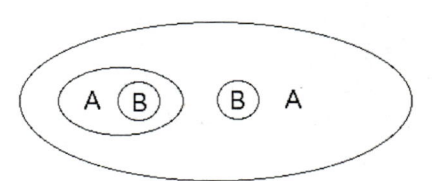

Which is equal to:

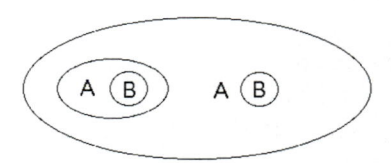

And this has the form:

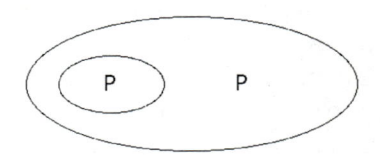

This is equivalent to:

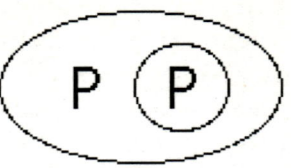

$$P(P)$$

which is always true since it expresses the truism "P implies P". In this way the existential graphs give access to the interior structure of the tautologies.

This is the mystery of elementary logic: *the interior structure of the tautologies.* What forms of utterance are necessarily true, which are contingent on circumstance and which are simply false? All logical systems aim at clarifying this matter. What is important about the existential graphs is that they allow the visual manipulation of complexes of Signs to arrive at the desired answers. A visual language for logic emerges from the existential graphs.

Extra decorations on the Existential Graphs allow them to include quantifiers and modal logic as well. What is quite fascinating in reading Peirce on these developments is his maintenance of a clear conceptual line connecting spoken and written language on the one hand, and diagrammatic and written formalisms on the other. The places where these domains can touch are sometimes sparse and delicate. A good example is implication: "not (A and not B)" is a denial in language drawing the precise boundary that defines implication. The logically equivalent statement "(not A) or B" is puzzling in ordinary language, requiring an analysis to ferret out its meaning. There is a subtle difference in the use of "or" as opposed to "and" in ordinary language. In ordinary language A "or" B usually means "A or B and not both". In standard logic it is "A or B or both" that is intended.

A minimal formalism may not be the most effective interface with speech and word, and yet the mathematician will continue to search for these least structures for the sake of economy, elegance and computational effectiveness. Peirce walks the creative middle road. Elegance and economy emerge nonetheless.

VI. Existential Graphs and Laws of Form

We now make a descent into the internal structure of the existential graphs in a direction that Peirce apparently did not take to its full conclusion. What is the truth value of the empty existential plane? In that plane nothing is asserted to exist. The truth value is True, T. An empty circle encloses empty space and so negates it, giving rise to the value False, F.

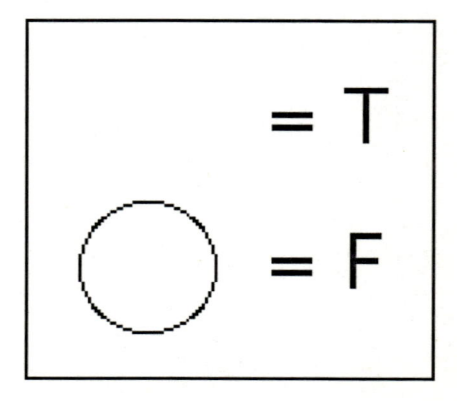

$$ = T$$
$$\bigcirc = F$$

The Mathematics of Charles Sanders Peirce

With this convention, we can evaluate patterns of adjacent and nested circles in the plane.

0. "nothing" = T
1. ( ) = F
2. False and False = False. Thus ( )( ) = ( ).

$$\bigcirc\bigcirc = \bigcirc$$

3. (( )) = (F) = not F = T. Thus (( )) = "nothing".

$$(\!(\,)\!) =$$

More complex expressions can be simplified uniquely by the successive application of these rules. For example

$$((\circ\circ)(\,))(\,) = (\,(\circ)(\,))(\,)$$

$$= (\,(\,)(\,)) = (\,) = \bigcirc$$

With this arithmetic of circles we can handle the evaluation of existential graphs by direct substitution. For example,

If P =

$$(P(P)) = (\,(\,)) = \;= T$$

If P = $\bigcirc$

$$(P(P)) = (\,(\circ)(\circ)\,)$$

$$= (\,(\circ)\,) = \;= T$$

Thus (P(P)) is true in all cases, as it should be, since (P(P)) expresses "P implies P" in the existential graphs.

Something else is going on here. While efficiently calculating the truth tables, the circles hold a simpler and wider meaning of their own. Each circle makes a distinction between its inside and its outside. It is this calculus of distinctions that handles the tautologies.

The two existential graphs are equivalent, exactly when they have the same circle evaluations for all possible substitutions of circles or blanks into the variables in the two graphs.

This arithmetic of circles, implicit in C.S. Peirce's existential graphs, is isomorphic with the primary arithmetic (calculus of indications) discovered by G. Spencer Brown in his lucid book " Laws of Form" [9].

The basic symbol in Laws of Form is a right angle bracket, rather than a circle, but its use is just the same (as an enclosure) and the primary algebra of Spencer-Brown is also isomorphic with the existential graphs themselves. In the usual interpretation for logic in Laws of Form one takes juxtaposition of forms to be "or" rather than "and" thus getting a dual calculus where (A)B stands for " A implies B" and () stands for "True".

Access to the primary arithmetic adds an extra dimension to the structure of the existential graphs.

This primary arithmetic of circles (or brackets) is a fundamental pattern underlying first order logic. First order logic is the mathematical pattern that emerges from the primary arithmetic. It requires some time to get used to this very different point of view about logic. We all know that logic is basically simple, and yet viewed from "and" and "or" and "implies" it has a curious complexity that baffles the intuition. Yet logic is nothing more than the properties of the act of distinction! At the level of the primary arithmetic we are returned to this enlightened state. Then we have to work hard to reestablish connection with the complex world that has been left behind.

We can use the primary arithmetic to verify many graphical identities, just as we verified (P)P = () above by considering the different substitutions of P as marked or unmarked.

Here is a list of some basic identities that can be checked in this fashion:
1. PP = P
2. () P = ()
3. ((P)) = P
4. (P)Q = (PQ)Q
5. (P(P)) =
6. ((P)(Q))R = ((PR)(QR))
7. ((P) (Q)) (P (Q)) = Q

With the help of these identities it is easy to decide on the validity of expressions in symbolic logic that are expressed in terms of the existential graphs. Identities 4. and 5. (plus the commutativity that is implicit in a planar representation) are sufficient to derive all valid equational identities in this system.

With more work one can in fact use identity number 7. as a basis for the logical algebra. To prove this was a difficult problem in the original context of Boolean algebra where no symbol could be unmarked. (In the usual notation the unmarking of a symbol could lead to unintelligible expressions such as a + '.) The problem was solved by Huntington in 1936 [19]. The question is easier when we allow an unmarked symbol. See [15], [16], [17], [18]. Since the seventh equation above is sufficient for the whole alpha theory of existential graphs, it is interesting to consider its interpretation in terms of illation (inference).

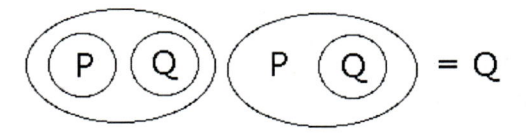

$$\left(\overline{(P)(Q)\;(\overline{P\;Q})}\right) = Q$$

This equivalence states that
"[[not P] implies Q] and [P implies Q]" is equivalent to "Q".
Once again, the tautology is understandable in the realm of ordinary language (it is an expression of the law of the excluded middle). It is remarkable that this single identity can be taken as the foundation for the theory of Peirce's alpha graphs.

It is important to note that with the primary arithmetic, Spencer-Brown was able to turn the epistemology around so that one could start with the concept of a distinction and work outwards to the patterns of first order logic. The importance of this is that the simplicity of the making (or imagining) of a distinction is always with us, in ordinary language and in formal systems. Once it is recognized that the elementary act of discrimination is at the basis of logic and mathematics, many of the puzzling enigmas of passing back and forth from formal to informal language are seen to be nothing more than the inevitable steps that occur in linking the simple and the complex. The elementary act, the deep structure, is not simple. The locutions of ordinary language are in fact quite simple but elaborate. These locutions enable us to speak without thinking. Yet they enable us to speak thoughtfully. It is in the articulation of careful thought that the descent into basic discrimination is called for without compromise.

In this view, the empty circle is not just a contracted or abstracted notation, but rather *an iconic representative of an elemental distinction*. The primary arithmetic is a mathematical language that is based on that distinction and incorporates it into its own symbol system.

In that view the identity ()() =()

$$\bigcirc\bigcirc = \bigcirc$$

states (when one circle is seen to be a copy or symbol for the other) the redundancy of naming the distinction with a copy of itself.

Here we make contact with the elements of Peirce's trichotomy

Sign   (Representamen)/Signified   (Object)/Interpretant.

The circle is a Sign for itself where "itself" is a distinction drawn in the plane and the Sign is also a circle drawn in the plane. The circle signifies a distinction and the distinction that is signified is made by the circle itself! Thus in the primary arithmetic or calculus of indications, we have a minimal form of that Peircian semiotic epistemology. The circle refers to itself, but that self is a Sign in a context of Signs, and so the circle can refer to other Signs individually indistinct from itself and yet distinguished from the original circle by the context of this community of Signs. In this community, an individual circle can appear or disappear, yet the identity of the circle as a Sign is inviolate. Here the interpretant may at first seem to be the formal system of the circles themselves, but this widens to include the person making these distinctions, and widens further to include the entire Sign complex that constitutes that person and is reflected in the arithmetic itself. All this remains true for numerical arithmetic and beyond. In the primary arithmetic the relation of Sign and distinction is transparent.

In Chapter 12 of Laws of Form [9], Spencer-Brown writes " We see now that the first distinction, the mark and the observer are not only interchangeable, but, in the form, identical. " The mathematician is not distinguished from the system that he/she is making.

In that view the identity (())= "nothing"

is interpreted as the statement

"A crossing from the marked state yields the unmarked state".

And the unwritten identity () = ()

is interpreted as the statement

"A crossing from the unmarked state yields the marked state".

In both cases the circle is viewed as either a noun (name of the outside of the distinction) or a verb (crossing from the state indicated on the inside of the circle). I would summarize what we have just said by the following sentence.

> In descending to the primary arithmetic, one enters a natural world of (formal) speech that has its own meaning in relation to a distinction (made by that speech itself), a meaning that informs and underlies the logic of language and mathematics.

This description marks the beginning of seeing Peirce's existential graphs in this light.

Remark on Notation.
We remark that in Spencer-Brown's Laws of Form [9], the notation for an enclosure is not a circle, but a right angle bracket.

Thus the laws of calling and crossing as we have drawn them in circles become the following patterns in the right angle bracket:

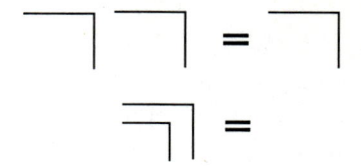

As we go from arithmetic to algebra and logic, Spencer-Brown makes the choice that AB (A juxtaposed with B) represents "A or B" rather than "A and B" as we have seen in the existential graphs and with the Sign of illation. However, with the marked state interpreted as "True" and the unmarked state interpreted as "False", implication in the Spencer-Brown algebra is given by the form shown below.

$$\overline{A}|B = \text{"A implies B"}$$

This puts implication in Laws of Form in exactly the same pattern as in Peirce's Sign of illation! In fact, now we have the following curious rogues gallery of notations for implication:

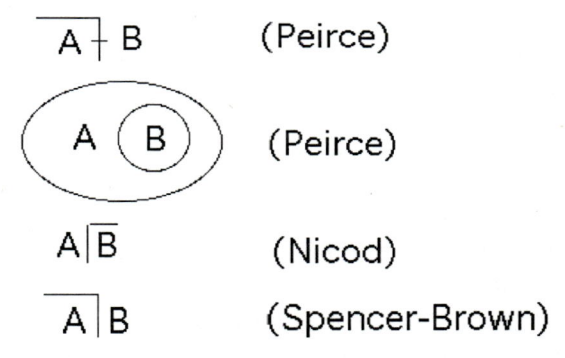

$A \vdash B$ (Peirce)

(A (B))   (Peirce)

$A|\overline{B}$ (Nicod)

$\overline{A}|B$ (Spencer-Brown)

The first Peirce notation is the portmanteau Sign of illation. The second Peirce notation is the form of implication in the existential graphs. The Nicod notation is a portmanteau of the Sheffer stroke and an (overbar) negation sign. The Spencer-Brown notation is in line with the Peirce Sign of illation. It remained for Spencer-Brown (some fifty years after Peirce and Nicod) to see the relevance of an arithmetic of forms underlying his notation and thus putting the final touch on a development that, from a broad perspective, looks like the world mind doing its best to remember the significant patterns that join logic, speech and mathematics. The movement downward to the Form ("we take the form of distinction for the form."[9, Chapter 1, page 1]) through the joining together of words into archetypal portmanteau Signs can be no accident in this process of return to the beginning.

VII. The Logical Garnet

The purpose of this section is to point out a remarkable connection between Laws of Form, the Existential Graphs of Peirce, polyhedral geometry, mirror symmetry and the work of Shea Zellweger [30].

Zellweger did an extensive study of the sixteen binary connectives in Boolean logic ("and", "or" and their relatives — all the Boolean functions of two variables), starting from Peirce's own study of these patterns. He discovered a host of iconic notations for the connectives and a way to map them and their symmetries to the vertices of a four dimensional cube and to a three dimensional projection of that cube in the form of a rhombic dodecahedron. Symmetries of the connectives become, for Zellweger, mirror symmetries in planes perpendicular to the axes of the rhombic dodecahedron. See Figure 2. Zellweger uses his own iconic notations for the connectives to label the rhombic dodecahedron, which he calls the "Logical Garnet".

This is a remarkable connection of polyhedral geometry with basic logic. The meaning and application of this connection is yet to be fully appreciated. It is a significant linkage of domains. On the one hand, we have logic embedded in everyday speech. One does not expect to find direct connections of the structure of logical speech with the symmetries of Euclidean Geometry. It is the surprise of this connection that appeals to the intuition. Logic and reasoning are properties of language/mind in action. Geometry and symmetry are part of the mindset that would discover eternal forms and grasp the world as a whole. To find, by going to the source of logic, that we build simultaneously a world of reason and a world of geometry incites a vision of the full combination of the temporal and the eternal, a unification of action and contemplation. The relationship of logic and geometry demands a deep investigation. This investigation is in its infancy.

In this section I will exhibit a version of the Logical Garnet (Figure 2) that is labeled so that each label is an explicit function of the two Boolean variables A and B. A list of these functions is given in Figure 1. We will find a new symmetry between the Marked and Unmarked states in this representation. In this new symmetry the mirror is a Looking Glass that has Peirce on one side and Spencer-Brown on the other!

Before embarking on Figure 1, I suggest that the reader look directly at Figure 2. That Figure is a depiction of the Logical Garnet. Note the big dichotomies across the opposite vertices. These are the oppositions between Marked and Unmarked states, the opposition between A and not A, and the opposition between B and not B. If you draw a straight line through any pair of these oppositions and consider the reflection in the plane perpendicular to this straight line, you will see one of the three basic symmetries of the connectives. Along the A/not A axis the labels and their reflections change by a cross around the letter A. Along the B/not B axis, the labels change by a cross around the letter B. These reflections correspond to negating A or B respectively. Along the Marked/Unmarked axis, the symmetry is a bit more subtle. You will note the corresponding formulas differ by a cross around the whole formula and that both variables have been negated (crossed). Each mirror plane performs the corresponding symmetry through reflection. In the very center of the Garnet is a double labeled cube, labeled with the symbols "A S B" and "A Z B". These stand for "Exclusive Or" and its negation. We shall see why S and Z have a special combined symmetry under these operations. The rest of this section provides the extra details of the discussion.

Let us summarize. View Figure 2. Note that in this three-dimensional figure of the Logical Garnet there are three planes across which one can make a reflection symmetry. Reflection in a horizontal plane has the effect of changing B to its crossed form in all expressions. Reflection in a vertical plane that is transverse to projection plane of the drawing, interchanges A and its crossed form. Finally, reflection in a plane parallel to the projection plane of the drawing interchanges marks with unmarks. We call this the Marked/Unmarked symmetry.

On first pass, the reader may wish to view Figure 2 directly, think on the theme of the relationship of logic and geometry, and continue into the next section. The reader who wishes to see the precise and simple way that the geometry and logic fit together should read the rest of this section in detail.

Figure 1 is a list of the sixteen binary connectives given in the notation of Laws of Form. Each entry is a Boolean function of two variables. In the first row we find the two constant functions, one taking both A and B to the marked state, and one taking both A and B to the unmarked state (indicated by a dot). In row two are the functions that ignore either A or B. The remaining rows have the functions that depend upon both A and B. The reader can verify that these are all of the possible Boolean functions of two variables. The somewhat complicated looking functions in the last row are "Exclusive Or", A S B and its negation A Z B.

In order to discuss these functions in the text, and in order to discriminate between the Existential Graphs and the Laws of Form notations, I will write <A> for the Laws of Form mark around A. Thus, in contrast, (A) denotes the Existential graph consisting in a circle around A. The Spencer-Brown mark itself is denoted by <>, while the circle in the Peirce graphs is denoted by (). Exclusive Or and its negation are given by the formulas:

$$A\ S\ B = <<A>B><A<B>>$$

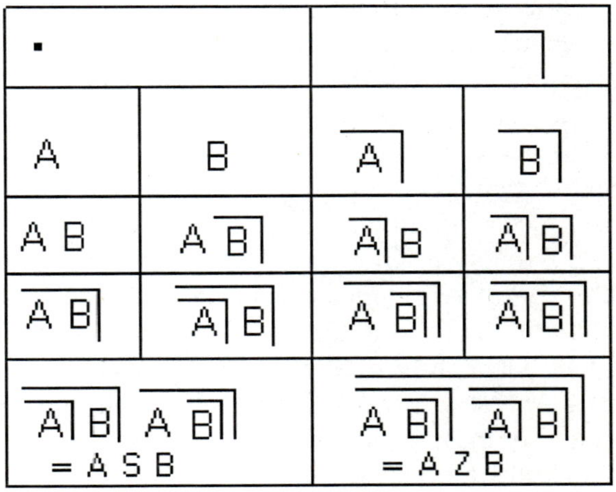

Figure 1. The Sixteen Binary Boolean Connectives

Here we have used Zellweger's alphabetic iconics for Exclusive Or with the letters S and Z topological mirror images of each other.

Exclusive Or is actually the simplest of the binary connectives, even though it looks complex in the chart in Figure 1. Let "Light" denote the unmarked state and "Dark" denote the marked state. Then the operation of Exclusive Or is as follows:

Dark *S* Dark = Light,
Dark *S* Light = Dark,
Light *S* Dark = Dark,
Light *S* Light = Light.

Imagine two dark regions, partially superimposed upon one another. Where they overlap, the darknesses cancel each other, and a light region appears. This is the action of Exclusive Or. Darkness upon darkness yields light, while darkness can quench the light, and light combined with light is light. In other words, Exclusive Or is the connective closest to the simple act of distinction itself, and it is closest to the mythologies of creation of the world (heaven and earth, darkness and light) than the more complex movements of "and" and "or". Exclusive Or and its negation sit at the center of the logical garnet, unmoved by the symmetries that interchange the other connectives.

The operation of Exclusive Or on the marked state is the same as negation (darkness cancels darkness to light) and the operation of Exclusive Or on the unmarked state is the identity operation that makes no change.

$$A\ S <> \ = \ <A>$$
while
$$A\ S\ .\ = A$$

The symmetries of Exclusive Or are very simple. If we change one of the variables to its negation we just switch from S to Z! That is,

$$<A> S\ B = A\ S <B>\ = A\ Z\ B.$$

As a result, A S B and A Z B together are invariant under the symmetries induced by the A/not A and B/ not B polarity.

We now discuss the Marked/Unmarked symmetry in the Garnet. This symmetry is available on a look at the Logical Garnet (Figure 2). For reference I have also shown the corresponding terms below (Diagram 1). Note that the horizontal lines in the middle of each diagram are not edges in the Garnet, but they do connect terms that correspond to one another under the mirror symmetry.

We see from Diagram one that the correspondence of terms in the Marked/Unmarked symmetry on the Logical Garnet is exactly the translation that we have described in this paper between the Laws of Form representation and Peirce's Existential Graphs! This last symmetry is actually a translation between two closely related languages for basic logic. This translation is a translation that interchanges "and" and "or". In fact, the reader will note that each of these terms occurs in between two single letters (possibly with one or both negated) on the Garnet. We have arranged the labels on the Garnet so that the front plane compound terms are the "or" of the adjacent vertices and the back plane terms are the "and" of the adjacent vertices.

Diagram 1 - Marked - Unmarked Symmetry

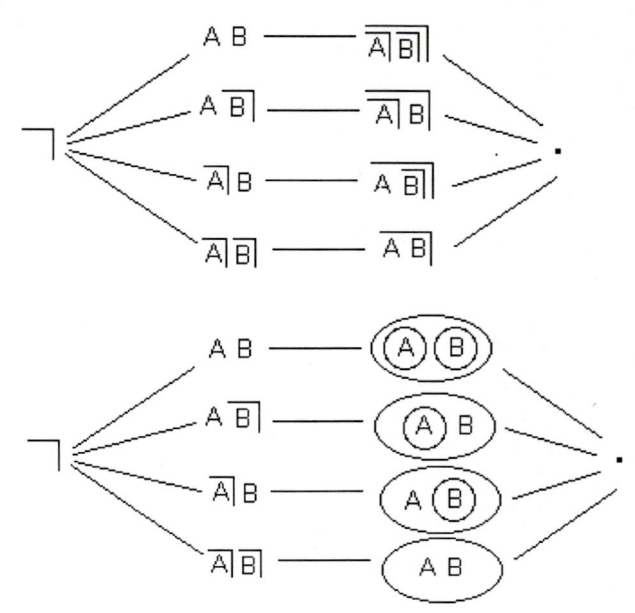

Note also that the central vertex (cube) in the Garnet (labeled with A *S* B and A *Z* B) is connected to the eight compound terms on the periphery of the Garnet. These terms are the terms that arise from Exclusive Or and its Complement when we take it apart. For example

$$A \; S \; B = <<A>B><A<B>>$$

and we can take this apart into the two terms

$$<<A>B> \text{ and } <A<B>>,$$

while

$$<A> \; S \; B = <AB><<A><B>>$$

and we can take this apart into the two terms

$$<AB> \text{ and } <<A><B>>$$

The reader will enjoy looking at the geometry of the way the central and simple operation of Exclusive Or is taken apart into the more complex versions of "and" and "or" and how the Geometry holds all these patterns together.

As for the periphery of the Garnet, it is useful to diagram this as a plane graph with the corresponding labels shown upon it. The illustration below (diagram 2) exhibits this graph of the rhombic dodecahedron.

The rhombic dodecahedron itself does not have a central vertex and the graph below shows precisely the actual vertices of the rhombic dodecahedron and their labels. By comparing with Figure 2, one can see how to bring this graph back into the third dimension. Note how we have all the symmetries apparent in this planar version of the rhombic dodecahedron, but not yet given by space reflection. It requires bringing this graph up into space to realize all its symmetries in geometry.

Diagram 2 - Planar Graph of the Rhombic Dodecahedron

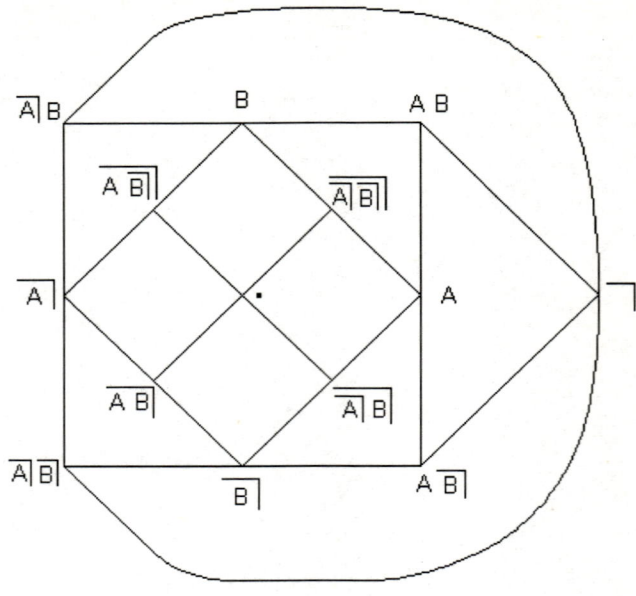

Diagram 3 - The Rhombic Dodecahedron as the Lattice Associated with a Two Circle Venn Diagram

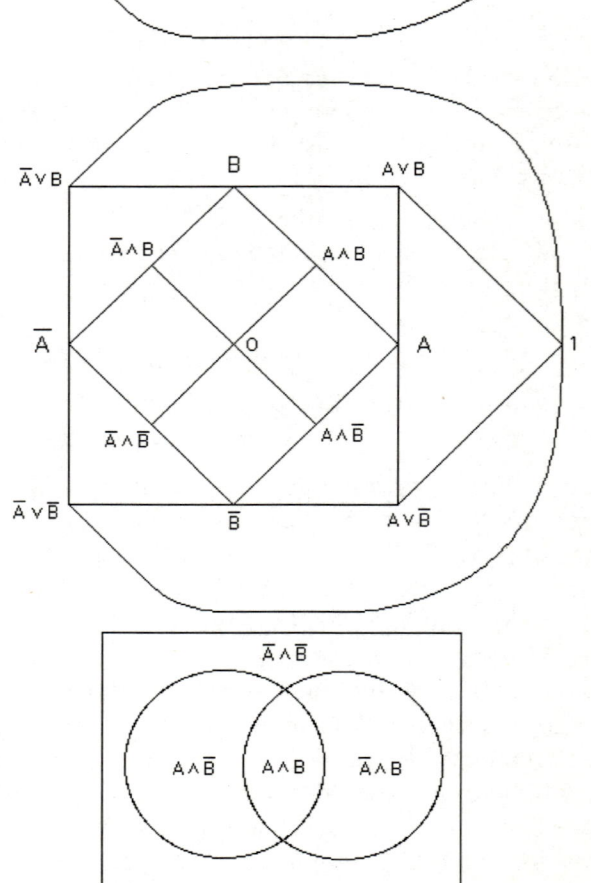

The Mathematics of Charles Sanders Peirce

Looking at this peripheral structure, we see the genesis of the pattern of the rhombic dodecahedron in relation to the connectives.

This graphical pattern can be viewed as the lattice of inclusions of these functions regarded as subsets of a universal set. To see this clearly, view the next diagram where we have labeled the vertices of the graph in standard notation with an upward pointing wedge denoting intersection ("and"), a downward pointing wedge denoting union ("or"), 0 denoting empty set and 1 denoting the universe. Then, going outward from 0, pairs of vertices are connected to vertices denoting the union of their labels until we reach the whole universe which is denoted by 1. This lattice is exactly the graph of the rhombic dodecahedron (Diagram 3).

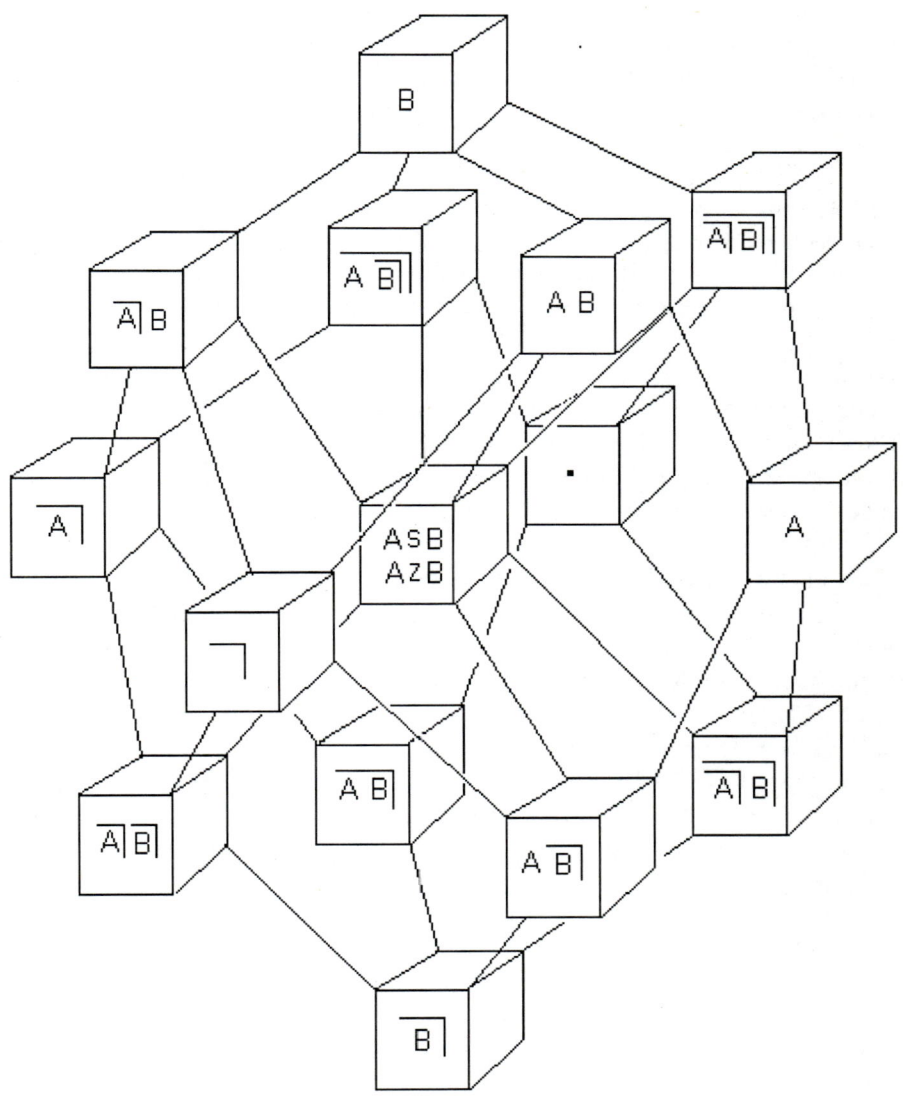

Figure 2 - The Logical Garnet

In this section we have exhibited a version of the Logical Garnet that intermediates between Peirce's Existential Graphs and the dual approach of Laws of Form. This appearance of significant Geometry at the very beginning of Logic deserves deeper investigation. The diagrammatic investigations of Peirce, Venn, Carroll, Nicod and Spencer-Brown are all ways of finding geometry in logic, but in Zellweger's Logical Garnet classical three-dimensional geometry appears, and this is an indication that one should think again on the relationship of logic and mathematics.

VIII. Infinity and Infinitesimals, Recursive Domains

Peirce was an advocate of the notion that infinitesimal numbers were as natural as the concepts of infinity and infinite numbers.

That there was a controversy over this point is a consequence of the history of the calculus where, at first, Newton and Leibniz both used infinitesimals freely. Later as a critical period set in, mathematicians decided to keep track of all approximations as precisely as possible and the concept of limits was born. With this, a direct need for infinitesimals vanished and the machinery of mathematical scholarship kept these "ghosts of departed quantities" in the background. Peirce was one of the few mathematical people who advocated infinitesimals in the early part of the twentieth century.

The situation did not begin to clear up until the 1970's when the logician Abraham Robinson [20] published his beautiful work showing how to work with infinitesimals in their full subtlety. Later developments produced different models of numbers that included infinitesimals with less formal machinery than the Robinson theory [20]. For example, there are the surreal numbers of John Conway [21], the square zero infinitesimals of Lawvere [27] and Bell [22], the sequence infinitesimals of Henle [23]. It will help this discussion to consider the concept of infinitesimal in an informal way, and then to compare with what Peirce said about them.

We imagine a new sort of positive number d that is not zero, and is nevertheless "smaller" than any ordinary positive number that you can name. This infinitesimal d is in itself a generator of other infinitesimals. Thus $d+d = 2d$ is also infinitesimal and larger than d, while $d \times d = d^2$ is smaller than d. If we take the reciprocal $1/d$ we obtain a number that is "larger" than any standard number. This means that $1/d$ is a kind of "infinite number", but $1/d$ is not the same as the reciprocal of 0 (and we do not allow $1/0$ in our calculations since it leads to contradictions). The addition of d and its powers and reciprocals to the number system actually does not , if handled correctly, lead to any contradictions. Adding d to the numbers is quite analogous to extending the number system to include the square root of minus one (to create the complex numbers). The extension of numbers to include infinitesimals can be accomplished, and once it is done, one can do calculus without using limits at the fundamental level. The basic idea is that with an infinitesimal one can study how a function changes "instantaneously". That is we can form the ratio (f(x+d)-f(x))/d and call the standard part (the non-infinitesimal part) of this quotient the *derivative* of the function f at the point x.

For example, if f(x) = xx (the product of x with itself), then

(f(x+d)-f(x))/d = ((x+d)(x+d) −xx)/d = (2xd − dd)/d = 2x −d.

Since 2x is the non-infinitesimal part of this difference quotient, we conclude that the derivative of xx is 2x.

The concept of infinitesimal is closely tied with the concept of continuity. Infinitesimals seem to form a glue that holds the points of the line together. These sorts of intuitions were at the core of Peirce's discussion of infinitesimals. Here is his voice:

> It is singular that nobody objects to the square root of minus one as involving any contradictions, nor, since Cantor, are infinitely great quantities much objected to, but still the antique prejudice against infinitesimally small quantities remains. [6, Vol. 3, p. 123].

A little later he continues with arguments relating this to our understanding of consciousness.

> It is difficult to explain the fact of memory and our apparently perceiving the flow of time, unless we suppose immediate consciousness to extend beyond a single instant. Yet if we make such a supposition we fall into grave difficulties unless we suppose the time of which we are immediately conscious to be strictly infinitesimal. [6, Vol. 3, p. 124]

In this way, Peirce identifies the infinitesimal with the consciousness of the immediate moment.

> We are conscious only of the present time, which is an instant, if there be any such thing as an instant. But in the present we are conscious of the flow of time. There is no flow in an instant. Hence the present is not an instant. [6, Vol. 3, p. 126]

By taking the stance that that there can be no movement in an instant, Peirce argues that the present (infinitesimal) moment cannot be an instant. Along with this argument for the notion that the perception of the present is not a point but rather an infinitesimal, Peirce takes the stance that the continuum of the line is not made of points and that any attempt to analyze the line into points will lead to higher and higher orders of infinity for the number of points on that line. This idea is precisely born out in the surreal numbers of John Conway [21]. See also discussion along a similar vein in the paper by Robin Robertson [34] in this volume.

It is interesting to speculate whether Peirce took these ideas of infinitesimals, continuity and infinity into the arena of his existential graphs. If so, he might have considered infinite graphs such as the one shown below.

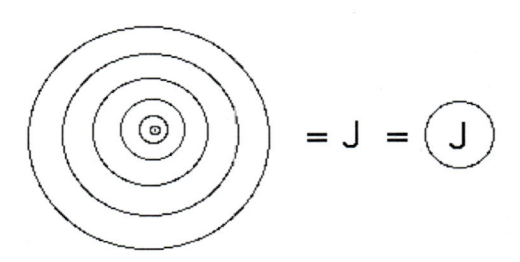

Peirce's view of an inexhaustible infinity is closely related conceptually with the reflexivity embodied in the self-containing form J depicted above. If we say that J is identical to J with a circle around it, then this identification is really the step of adding one more circle to the pattern. It is a conceptual recognition of the potential endlessness of the series of nested circles. Just as Peirce's work with the existential graphs skirted close to the underlying structure of the primary arithmetic of the circles themselves, so does his work on continuity skirt close to the paradoxical and topological patterns of reflexivity. These patterns have flowered in modern logic through the work of Church and Curry [25] (lambda calculus), Gödel [31], [32], [33] (self-reference and incompleteness), and Turing [33] (recursive instruction and ideal computers). In the tradition of existential graphs we might write a new graph just designed to indicate this reflexive reentry of the form into itself that takes place not in an instant but in a time rendered infinitesimally small.

$$J = \text{(figure)} = \text{(figure)}$$

While it is certainly speculation to imagine that Peirce entertained the formalisms of self-reference and re-entry, nevertheless the context for such constructions is indeed not far from his point of view about the nature of mathematics. It is clear from his writings that he regarded mathematics as a creative enterprise where one could make a hypothesis, draw a figure, visualize a pattern and follow out the consequences of this activity. If it should happen that the assumptions lead to a contradiction, then that is a result in itself. Thus he would take infinitesimals as innocent before proven guilty. And in this way, he managed to foretell the fate of these structures as they were indeed pronounced innocent by the court of Abraham Robinson [20].

This human and constructive attitude (innocent before proven guilty as in the last paragraph) toward mathematics would have made Peirce quite receptive to the reflexive domains of the untyped lambda calculus [25] where the fixed points and self-references can exist. A lambda domain in this sense is a class of objects that can act on one another to form new objects of the same kind.

The action of x on y is denoted by the juxtaposition xy. The characteristic of a lambda domain is that functional operations on the domain acquire names and become objects in the domain. Thus if we decide that F will operate on the domain via Fx=(xy)x for some fixed y, then this definition of F is sufficient to allow it membership in the domain. A lambda domain is like a computer language where you can add new words to the language by defining them as actions on previously created words, and allowing these actions to extend to the new words themselves so that a word can act on itself. This dictum has recursive consequences as we shall see in a moment, but my point is that Peirce's view of mathematics as a whole really is that Mathematics is a lambda domain managed (gardened) by the judgment of the human beings who sign, signify and make references in that garden. This means that the function of the mathematician is not to determine the eternal nature of the objects in the garden, but rather to find that they and the mathematician himself (or herself) are all Signs, growing together in the expansion of Language.

The reflexive and recursive nature of lambda domains was recognized most clearly by Church and Curry [25] who proved the

>**Theorem**. For every element **F** in a lambda domain, there is a **J** in that domain such that **F(J) = J**.

This is the *Fixed Point Theorem of Church and Curry*.

>**Proof.** Let **G** be defined by **Gx = F(xx)** .
>**G** itself is in the lambda domain. Thus we can form **GG**.
>But **GG = F(GG)** since **Gx=F(xx)** for any **x**.
>Therefore take **J=GG** and conclude that **J=F(J)** as desired. //

This fixed point theorem is in the very center of modern logic, mathematics and computer science. It encodes most of the known paradoxes and the form of Gödel's Incompleteness Theorem, the essence of recursion and the sizes of infinity [27] in a guise of extreme simplicity. Peirce would have approved.

It would take us too far afield to mark out just how this one fixed point theorem is the core of so many apparently diverse matters. Some of these themes have already been taken up in the author's columns for Cybernetics and Human Knowing [35]. The article [27] by Lawvere is also a useful introduction. The relation with paradox is easy and will give the flavor of the matter: Let **F** be the operation of negation ~. Then the Theorem supplies us with J such that ~**J** = **J**. An entity that is equal to its own negation is the same as an entity that asserts "I am a liar." Thus the famous Paradox of the Liar is a direct production of the Fixed Point Theorem.

Lambda domains have been formalized and partially tamed for the sake of mathematical foundations and computer science. Philosophical biologists and cyberneticians such as F. Varela [10], H. Maturana and F. Varela [11] L. H. Kauffman and F. Varela [12], Heinz von Foerster [13] and L. H. Kauffman [14] have written eloquently of the basic nature of reflexive domains in relation to the biological and linguistic view of Nature inseparable from her sentient creations.

Leibniz, seventeenth century philosopher, logician and mathematician dreamed of a logical calculus of thought that would allow persons inclined to investigate a topic or settle a dispute to simply sit down and calculate together to come to agreement and knowledge. The active possibility of a symbolic logic with the power to encompass Leibniz's dream of a *calculus ratiocinator* was present to Peirce and the logicians of his generation with many hoping to create that language in thought and diagrams. The dream has now expanded into the present world of recursive complexity. The dream has not disappeared.

IX. A Sign of Itself

There is clearly much more to be done in this arena of investigation and speculation into the nature and structure of the mathematics of Charles Sanders Peirce. I believe that the key to understanding Peirce on mathematics is his view of the nature of a human being as a Sign. In this view there can be no essential separation of the human being and the mathematics or language of that being. This may seem a radical stance in this antiseptic age. Here is Peirce himself, speaking about "a Sign of itself":

> "But in order that anything should be a Sign it must 'represent', as we say, something else called its *Object,* although the condition that a Sign must be other than its Object is perhaps arbitrary, since, if we insist upon it we must at least make an exception in the case of a Sign that is part of a Sign. Thus nothing prevents an actor who acts a character in a an historical drama from carrying as a theatrical 'property' the very relic that article is supposed merely to represent, such as the crucifix that Bulwer's Richelieu holds up with such an effort in his defiance. On a map of an island laid down upon the soil of that island there must, under all ordinary circumstances, be some position, some point, marked or not, that represents *qua* place on the map the very same point *qua* place on the island...
>
> If a Sign is other than its Object, there must exist, either in thought or in expression, some explanation or argument or other context, showing how – upon what system or for what reason the Sign represents the Object or set of Objects that it does. Now the Sign and the explanation make up another Sign, and since the explanation will be a Sign, it will probably require an additional explanation, which taken together with the already enlarged Sign will make up a still larger Sign; and proceeding in the same way we shall, or should ultimately reach a Sign of itself, containing its own explanation and those of all its significant parts; and according to this explanation each such part has some other part as its Object." [24]

In this passage Peirce speaks as a topologist. He tells us that if we overlay or in any (continuous) way place the map of a territory upon that territory then there must be a point on the map that coincides with the corresponding point on the territory. At first this statement might seem quite astonishing, but it is indeed true and it is the content of the famous

Brouwer Fixed Point Theorem. If **F** is any continuous mapping of a disk **D** to itself, then there exists a point **p** in the disk **D** that is left fixed by the mapping: **F(p) =p**.

The proof of this Theorem is illuminating and we refer the reader to its exposition in [26]. Presumably Peirce is assuming that his territory is in the topological shape of a disk. Otherwise the result is false. Imagine a world in the shape of a donut and a map of that world just the size of the world itself. Rotate the map a small amount in both of the turns available on a donut and every point of the map will move away from itself. In this example of a world in the shape of a donut, there are mappings that do not have fixed points.

In a universe in the shape of a disk, let the map be of the same size and shape as the disk itself. If we rotate that map about the center point of the disk by a small angle, then the center point of the map will coincide with the center point of the territory, and this will be the only point with this property.

But of course Peirce is not just assuming that the territory mapped is in the shape of a disk. He is using topology as an amplifier for our thought about self-reference. We are all familiar with the situation of superimposing a map on its territory. We have all been in a park and encountered a map with a marked point signifying "You are here." We have all seen that the orientation of the map itself in the space may not match the actual directions, but the truth of the self-locator on the map is still most useful.

In fact, Peirce in this passage is coming very close to the message of the Fixed Point Theorem of Church and Curry that we discussed in the last section. When he says of the place of coincidence of map and territory "we shall reach, or should, ultimately reach a Sign of itself, containing its own explanation and those of all its significant parts: and according to this explanation each such part has some other part as its Object" he is describing a Sign that refers to a significant part of itself and through that to itself. The Sign should "contain its own explanation". This is the reflexive or recursive nature of the reentrant or self-referential form.

Compare this discussion with the reentrant J of the previous section. The equation

$$J = \boxed{J}$$

asserts the reentry of J into its own indicational space, and it exhibits J as a "part of itself". The equation is the explanation of the nature of J as reentrant and can be taken as a description of the recursive process that generates an infinite nest of circles. It is only *J as an equation* that yields J as a Sign of itself. If we wish to embody the equation in the Sign itself then we need to allow the Sign to indicate its own reentry as we did in the last section with the symbol shown below.

This symbol does "contain its own explanation" in the sense that we interpret the arrow as an instruction to reenter the form inside the circle (ad infinitum). Self-reference is infinity in finite guise.

It has been said that "the map is not the territory" and this is indeed correct. But the most interesting terrain is that territory where we have no choice but to use the territory in the course of the construction of the map. And this is exactly what is done in mathematics, linguistics and science. In order to study language one must use language. In order to study mathematics one must use mathematics, and indeed we use mathematics to elucidate mathematics. Map and territory grow and evolve together in the course of time. In this view it is obvious that any attempt to fully explain anything will cause the map and territory to expand into a new domain in which further explanation will be needed. As Spencer-Brown says ([9] p.106) "In this sense, in respect of its own information, the universe must expand to escape the telescopes through which we, who are it, are trying to capture it, which is us."

The lambda domain of the previous section is a territory that is susceptible to self-evolution. Functions and descriptions of the lambda domain are also elements of that domain. And even though a lambda domain is not a topological disk, there will (via the fixed point theorem) be necessary points of coincidence between the points in the territory and the descriptions (maps) of that territory. It is an abstract model of language as a texture not of just words, but speakers. Each speaker is both noun (person, listener) and verb (person, speaker). Each person is his/her own explanation, but that explanation is a function of the entire domain of language, including the explanation itself.

X. Peirce and Second Order Cybernetics

Peirce, in speaking of the necessary occurrence of a "Sign for itself" in the relation of map and territory is referring, through a topological metaphor, to the reflexive nature of the domain of human discourse. Here is the conduit between Peirce and second order cybernetics.

In the last part of the quoted passage in the last section Peirce speaks of a hierarchy of Signs and explanations leading eventually to the Sign of itself " containing its own explanation and those of all its significant parts." The hierarchy occurs each time one looks into the context for the explanation of the given Sign. Sign and explanation form a new Sign to be explained ad infinitum (or in a circular network of explanations of explanations).

In the case of the lambda calculus or the simple infinite nest of circles we see images of this process of enfoldment where a larger external context is kept in the background. It is important to realize the extent to which we will keep such a background hidden for our own convenience! "I am the one who says I" and indeed the Sign "I" is a Sign for itself, but the full context is the entire English language and all its speakers, each of whom says "I". In a restricted context, one may manage without being engulfed by the language as a whole, and this is indeed the game played by a mathematician (or Humpty Dumpty! [3]) who would have words mean what he wants them to mean in a special context. The cost to Humpty Dumpty is well known; the cost to the mathematician is the emergence of paradox and complexity.

To avoid paradox in the lambda calculus, just such a (restricted) hierarchy was constructed by Dana Scott [28]. Scott used topology and recursive construction to create a reflexive space where every homeomorphism (continuous self-mapping with continuous inverse) of the Scott Space corresponds to a point in the space itself. (See [29].) Unbeknownst (perhaps!) to Scott, Peirce had laid down the program for such a construction many years before in his theory of Signs. As we have seen before, Peirce is a presence in the background of Modern Logic.

In reflexivity and in second order cybernetics Signs and their Objects become inextricably interlinked. Here is how Peirce puts the matter:

According to this, every Sign has a *Precept* of explanation according to which it is understood to be a sort of emanation, so to speak, of its Object. (If the Sign be an Icon, a scholastic might say that the 'species' of the Object emanating from it found its matter in the Icon. If the Sign be an Index, we may think of it as a fragment torn away from the Object, the two in their Existence being one whole or a part of such a whole. If the Sign is a Symbol, we may think of it as embodying the 'ratio' or reason, of the Object that has emanated from it. These of course are mere figures of speech; but that does not render them useless.) [24]

Here Peirce speaks of the interlocking relationship of Sign and Object. An example of the use of this concept in mathematics is the notion of *Gödel numbering* where the Sign for a text is a code number assigned to that text (by a definite procedure specified beforehand). The text is the Object and its Indexical Sign is the Gödelian code number. The reason for the use of such coding is that it then becomes possible for sentences in a formal system to refer to themselves by referring to their own(!) code numbers. This form of controlled self-reference was used by Gödel to prove that sufficiently rich formal systems are either inconsistent or incomplete. His Theorem shows that mathematics can not be encompassed by any single formal system.

Here follows a sketch of how this self-reference is accomplished. In it we shall see an astonishing interlock of Sign and Object.

In the formal context of this Gödelian work, a text can have a free variable that refers to a numerical value. Thus we may write

$$g \rightarrow T(u)$$

denoting the text by $T(u)$ with its free variable u, and g is the Gödel number of the text. The interlocking relationship between Sign and Object is specifically the fact that the text can use numbers and these numbers can be code numbers for other texts. In particular one can define a function from code numbers to code numbers (from Signs to Signs) as follows

#g is equal to the Gödel number of the text T(g) (with g substituted for u) when g is the Gödel number of T(u).

Thus if

$$g \rightarrow T(u)$$

then

$$\#g \rightarrow T(g).$$

The movement from the Sign g to the Sign #g is a shift of reference in which the original "name" g is now inherent in the Object $T(g)$ and that Object $T(g)$ has acquired the new name #g. As Peirce says, "If the Sign be an Index, we may think of it as a fragment torn away from the Object, the two in their Existence being one whole or a part of such a whole." Here #g is the "fragment" (disguised by the coding method) of the Object $T(g)$ that holds g within it.

Once this formality of the Gödel numbering gets underway, it is possible to have the actual situation of a fragment in the sense that the text speaks directly about #G, and #G is the Sign of that text. Gödel's trick is to first consider a Sign and Object in the form

$$G \to T(\#u)$$

Here G is the Sign of the Object T(#u). If we apply the shift to this pair we obtain

$$\#G \to T(\#G)$$

and the Sign is indeed a fragment of the Object. Since the Object is itself a referential text discussing #G, this text discusses its own Sign.

It is through this interlock of Sign and Object that Gödel constructs a text that asserts its own unprovability in the given formal system. See [18].

The interlocking relationship of Sign and Object was already well understood by Peirce. The mathematical ingredient for Gödel is the careful use of a restricted context (the given formal system).

The Gödelian result is the ultimate inability of such restricted formal systems to express the full range of mathematical truth. The full Sign of itself lies beyond such restrictions. Just so, we (who are the embodiments of the Sign of itself) can prove the Gödelian Theorem.

It is at the level of the Sign of itself that the Theorem, unprovable in the restricted system, can be proved. The human level, transcending the level of the restricted formal system, is the level of a Sign that is a Sign for itself.

Let us not forget the circle.

As we saw in descending from Peirce's existential graphs to the calculus of indications (by allowing a variable to take the unmarked (true) state), the circle

$$\bigcirc$$

lives in a language where it is a sign of itself. That language, the calculus of indications, unfolds the patterns of the existential graphs and marks a larger unfolding of language, mathematics and logic as a patterning of possible distinctions.

This point has been discussed at greater length earlier in this paper. We bring it up again here to remind the reader of the essential reflexivity of the basic language in which each circle, seen as a distinction, refers to any other circle seen as a distinction. The two equations of the calculus of indications are each self-referential in this sense. In the first equation (see below) either circle can be regarded as the name of the other circle. In the second equation, one circle acts as instruction to cross from the marked state that is indicated by the other circle. All expressions in the calculus of indications are self-referential. The more objective forms of logic and communication are based on this ground of circularity!

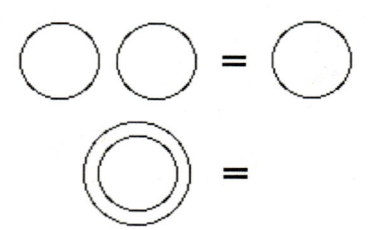

$$\bigcirc \bigcirc = \bigcirc$$

$$\circledcirc =$$

As a Sign of itself, the circle has only itself as a part. That part, equal to the whole, makes the distinction that is the referent of the Sign.

The explanation of this Sign is the Sign itself. The explanation of the circle is what it does in the plane upon which it is drawn. And that doing is the separation and joining of the inside of the circle with its outside. The circle is its own explanation. The circle is a Sign of itself that has no proper part.

And lest the tale of the circle seem too mystical or too particular, let us not forget that this circle is seen to do all that it does by an Observer who is the constructing of the circle.

The Observer is the Object that is the emanation of the Sign of Distinction.

The Distinction is the Object that is the emanation of the Sign of the Observer.

The act of drawing the circle is the motion of a point outward from an original location, only to return to that source in the primordial act of self-reference.

There is no plane, no circle but only an act that moves outward from self-identity and returns to identity.

That act is its own explanation.

XI. Epilogue

We have seen that Logic can pack a double meaning, that Logic could be an encoded form of Geometry. Peirce's portmanteau Sign has expanded to a vastness of multiple interpretations. This was implicit in Boole's original symbolic logic. He borrowed the symbolism of ordinary algebra and invited that symbolism to carry the structure of class and inference. The fit of a good portmanteau operates like a key in a lock, opening a connection between the apparently separate domains that compose it.

Another portmanteau lives in Logic. It is the Gödelian sentence that asserts its own unprovability. This sentence carries a double meaning. Inside the formal system, it is a statement about properties of certain integers. Within the formal system there is no hint that the Gödel sentence has any other meaning. From outside the formal system, the sentence is seen to assert its own unprovability. These two meanings interlock in the compound Sign that is the Gödel sentence, to form a portmanteau that has forever changed our understanding of the nature of formal systems. This understanding is already present in Peirce through his view of the nature of the perceiving consciousness as a Sign for itself.

Examine Charles Peirce's Sign of illation in the light of these reflections. The Sign lives in its own system of inference and needs no hint of the encoding that translates it to the world of Boolean algebra. The Sign has its own life in one language, and a description in terms of component parts in the other. Which is syntax and which is semantics? The portmanteau Sign is born of a linking of syntax and semantics - a portent of the future of language, logic and mathematics, as we have outlined in the body of this paper with the connections with Gödel's Theorem, lambda calculus and the calculus of indications of G. Spencer-Brown.

We ourselves are portmanteau Signs of a complex order. We are packing cases of multiple meaning large enough to make a human being a Sign of itself.

References

1. C. S. Peirce, "The New Elements of Mathematics", edited by Carolyn Eisele, Volume IV – Mathematical Philosophy, Chapter VI – The Logical Algebra of Boole. pp. 106-115. Mouton Publishers, The Hague – Paris and Humanities Press, Atlantic Highlands, N. J. (1976).
2. Webster's New Collegiate Dictionary, G. C. Merriam Co. Pub. Springfield, Mass. (1956).
3. Lewis Carroll, " Alice's Adventures in Wonderland & Through the Looking Glass", (1865), (1871), (1988) Bantam Books.
4. Lewis Carroll with notes by Martin Gardner, " The Annotated Alice - Alice's Adventures in Wonderland & Through the Looking Glass" New American Library (1960).
5. J. G. P. Nicod, A Reduction in the number of Primitive Propositions of Logic. Proc. of Cambridge Phil. Soc. Vol. 19 (1916), pp. 32 – 40.
6. C. S. Peirce, "The New Elements of Mathematics", edited by Carolyn Eisele, Volume III/1 – Mathematical Miscellanea, Lowell Lectures ,1903. Lecture II, pp. 406-446. Mouton Publishers, The Hague – Paris and Humanities Press, Atlantic Highlands, N. J. (1976)
7. K. L. Ketner, "Elements of Logic – An Introduction to Peirce's Existential Graphs" Texas Tech University Press (1990).
8. C. S. Peirce, "Collected Papers – IV Chapter 3 – Existential Graphs", pp. 4.397-4.417, edited by Charles Hartshorne and Paul Weiss, Harvard University Press, Cambridge (1933).
9. G. Spencer-Brown, "Laws of Form", Julian Press, New York (1969).
10. F. J. Varela, "Foundations of Biological Autonomy", North Holland Press (1979).
11. H. R. Maturana and F. J. Varela, "The Tree of Knowledge – The Biological Roots of Human Understanding", New Science Library (1987).
12. L. H. Kauffman and F. J. Varela, Form dynamics, J. Soc. and Biological Structures (1984).
13. H. von Foerster. "Observing Systems", Objects: Tokens for Eigenbehaviours, pp. 274 – 285. Intersystems Publications (1981).
14. L. H. Kauffman. Self-reference and recursive forms. J. Social and Biological Structures (1987), 53-72.
15. L. H. Kauffman. The Robbins Problem – Computer Proofs and Human Proofs. (to appear in the Festschrift in honor of Gordon Pask).
16. L. H. Kauffman. Imaginary values in mathematical logic. Proceedings of the Seventeenth International Conference on Multiple Valued Logic, May 26-28 (1987), IEEE Computer Society Press, 282-289.
17. L. H. Kauffman, (1990). Robbins Algebra. Proceedings of the Twentieth International Symposium on Multiple Valued Logic. 54-60, IEE Computer Society Press.
18. L. H. Kauffman. Knot Logic. In Knots and Applications ed. by L. Kauffman, World Scientific Pub. (1994), pp. 1-110.
19. Huntington,E.V. (1933) , Boolean Algebra. A Correction. *Trans. Amer. Math. Soc.* **35** , pp. 557-558.
20. A. Robinson, "Non-standard Analysis" (1966), North Holland – Amsterdam.
21. J. H. Conway, "On Numbers and Games" Academic Press (1976).
22. J. L. Bell, "A Primer of Infinitesimal Analysis", (1998) Cambridge University Press.
23. J. M. Henle, Non-nonstandard analysis: real infinitesimals, Mathematical Intelligencer , Vol. 21, No 1. (1999), pp. 67 – 73. Springer-Verlag, New York.
24. C. S. Peirce, "Collected Papers – II, p. 2.230 – 2.231, edited by Charles Hartshorne and Paul Weiss, Harvard University Press, Cambridge (1933).
25. H. P. Barendregt. "The Lambda Calculus Its Syntax and Semantics", North Holland (1981 and 1985).
26. R. Courant and H. Robbins, "What is Mathematics?", Oxford University Press (1941 and 1969).
27. F. W. Lawvere, Adjointness in foundations, Dialectica 23:82 (1969).
28. D. Scott, Continuous lattices, in "Toposes Algebraic Geometry and Logic", edited by F. W. Lawvere, pp. 97-136. Springer Verlag Lecture Notes in Mathematics Vol. 274 (1970).
29. S. Maclane and l. Moerdijk, "Sheaves in Geometry and Logic", Springer-Verlag (1992).
30. S. Zellweger, Untapped potential in Peirce's iconic notation for the sixteen binary connectives, in "Studies in the Logic of Charles Peirce", edited by N. Hauser, D.D. Roberts and J. V. Evra, Indiana University Press (1997), pp. 334-386.
31. D. Hofstadter, "Godel, Escher, Bach: An Eternal Golden Braid" Basic Books Inc. (1979).
32. E. Nagel and J. R. Newman, "Godel's Proof", New York University Press (1960).
33. J. N. Crossley et al, "What is Mathematical Logic?", Oxford University Press (1972).
34. R. Robertson, One, Two Three ... Continuity: C. S. Peirce and the Continuum, Cybernetics and Human Knowing (2001).
35. L. H. Kauffman, "Virtual Logic", columns in Cybernetics and Human Knowing.

Signs in Action:
Tarot as a Self-Organized System

Inna Semetsky[1]

Abstract: From a semiotic viewpoint, Tarot has been described as a mere artifact with pictorial cards being signifiers in a symbolic sense. This paper reconceptualizes the process-structure of Tarot by placing it in a three-fold framework that merges semiotics with systems-theoretical and cybernetic perspectives. Charles Sanders Peirce's triadic logic embedded in the action of signs, or semiosis in mind and nature, serves as a point of departure. By addressing Tarot from the position of general systems theory it is possible to describe Tarot dynamics by means of a sort of indexical connection to its signified. The latter, albeit functioning in a symbolic sense as an archetypal field of Jungian collective unconscious, is nevertheless capable of producing real effects at the level of human emotions, cognitions and habitual behaviors. The interpretation of symbols as the Peircean category of Thirdness creates, by virtue of mediation, a feedback loop, that is it generates conditions of possibility for self-organization. The previously unconscious, that is as yet out of conscious awareness, contents of one's mind become available to human knowing. The process of reading and interpretation contributes to, in a pragmatic sense, the creation of meanings for mental representations, the former inferred from the symbolism embedded in Tarot pictures. As such, Tarot as a self-organized system has the potential to provide epistemic access to the Peircean virtual Real, affirming, in a way, some contemporary debates of cognitive science.

It is honorable to be an epigone of Peirce. (Shimony,1993/II: 245.)

Introduction

The word *sign* is ambiguous. While traditionally defined as something that stands for something else, the notion of a sign as used in this paper follows Charles Sanders Peirce's triadic conception so as to underline the dynamic character of a sign. A sign can be anything that stands for something else, its object, in such a way so that to generate another sign, called its interpretant. In the broadest sense, Peirce used the word *representamen* to designate a sign, in agreement with the word *representation* describing the process, as well as the terminus of such a process, by which one thing stands for another. Each representamen is related to three things, the *ground,* the *object* and the *interpretant.* With respect to its ground, the representamen, as "used by every scientific intelligence ..., may embody any *meaning"* (Peirce CP 2. 229).

With regard to the latter, and in the spirit of Peirce's triadic semiotics, the structure of this paper will be three-fold. First, I will describe Peirce's logic and his pragmatism as a method of ascertaining the meaning of an idea. Second, I will

---

[1] Program in the Philosophy of Education, Columbia University Teachers College, New York, USA. Email: irs5@columbia.edu

turn to Carl Jung's depth psychology, with the purpose of emphasizing the pragmatic character of his analytical method, as well as the nature of archetypal patterns as analogous to Peircean habits in terms of being a psychological *ground* for the latter.

Third, I will address the structure and dynamics of Tarot as a self-organizing system of signs, drawing from both Peirce and Jung and asserting, in turn, a tarot reading as consistent with pragmatic method by means of creating a field of *meanings* for its interpreter rather than discovering some abstract universal, capital-T, Truth. Such an approach agrees in principle with the view that regards C.S. Peirce as one of the founders of constructive post-positivist philosophy (see Griffin 1993).

On Peirce's Triadic Logic of Relations.

The modern conception of logic has been developed by Peirce to include a general theory of signs, making semiotics tantamount to logic. While representing, in a narrow sense, the necessary conditions for the attainment of truth, logic for Peirce "is a science of the necessary laws of thought, or, better still (thought always taking place by means of signs), it is a general semeiotics, treating not merely of truth, but also of the general conditions of signs being signs" (Peirce CP 1. 444). Peirce also includes in his conceptualization the ideas of the functionally indubitable sets of beliefs that, as habits of human conduct, are not only culturally produced but are derived from a common layer of experience shared by all humankind.

Peirce's pragmatism, as such, blends logic and psychology and allows for the presensory and preconscious – not limited to sense-data – apprehension of reality upon which, despite its being necessarily vague, people are prepared to act, thus making their privileged beliefs presupposed in practice. Objective logic, for Peirce, is based on speculative grammar, the function of which is to provide the nature of the sign. The action of signs, or semiosis, is constituted by the relationship between an object and a mind by virtue of a sign, so that a sign is both affected by the object and is affecting the mind thus producing an effect – or meaning – called by Peirce the interpretant of the sign. The object to which the sign refers may not have a solely physical existence but may as well be a thought, a dream or an imaginary entity.

The triadic nature of relations between signs leads to Peirce's classifying signs in terms of basic categories of Firstness, Secondness and Thirdness: "First is the conception of being or existing independent of anything else. Second is the conception of being relative to, the conception of reaction with, something else. Third is the conception of mediation, whereby first and second are brought into relation.... In psychology Feeling is First, Sense of reaction Second, General conception Third, or mediation. ... Chance is First, Law is Second, the tendency to take habits is Third. Mind is First, Matter is Second, Evolution is Third" (Peirce CP 6.7). Firstness is quality, potentiality, freedom. Secondness, as a relation of the

First to the Second, is of opposites, physical reality, billiard-ball forces, rigid deterministic laws, direct effect, action and reaction. Thirdness relates seconds to thirds; it is synthesis, communication, memory, mediation.

A relational typology of semiosis – or, in other words, its process-structure – includes a sign, an object and an interpretant. The semantic criterion of sign-object relation includes icons, indexes and symbols; all three, depending on the pragmatic function of each, are liable to mutually exchange their roles. The icon is a sign which is capable of signifying by its own quality, the index is in some way dependent on its object, and the symbol, as saturated with significance, usually designates a conventional sign. Examples of icons include pictures and diagrams to the extent that they signify by virtue of some relative and relational similitude between the sign and what it stands for. Iconicity in turn can be further classified into three categories, the first being hypoicons that, according to Peirce, include both images and metaphors, or even a "pure fiction" (Peirce CP 4. 351) as objects of an icon.

The causal influence embedded in the semiotic process of cognition becomes indirect and moderated by means of inclusion of the third category that breaks down the direct dyadic cause-effect connection. Nonetheless the formal, albeit vague, principle, called by Peirce the rule of abduction, enables mind to reason from the premise to the conclusion; such an inference being described by the following statement: if A is B, and C can be signified by B, then maybe A is a sign of C. The interpretation is triggered by the Firstness of abduction which, functioning as a sort of perceptual judgement, is a hypothesis-bearing statement that asserts its conclusion only conjecturally; yet, according to Peirce (CP 5. 189), there is a reason to believe that the resulting judgement is true.

The given premise must entail some empirical consequences; the explication of the initial perception is achieved by analogical reasoning which unfolds into inferences to the would-be consequences of abductive conclusions eventually leading "to a result indefinitely approximating to the truth in the long run" (Peirce CP 2. 781), merging into synthetic inference in the process. The epistemic process, for Peirce, means denial of the Cartesian notion of arriving at propositions that mirror reality. The whole notion of a proposition, whose subject designates reality and whose predicate describes the essence of the said reality, is transformed by Peirce into interpretation of reality and living it out experientially: mimesis turns into semiosis.

Indeed, this is a sign by knowing which we know something more. Pragmatism as a method of ascertaining the meaning of ideas, understood by Peirce as intellectual concepts, the latter having been enriched with the qualitative Firstness of affects and emotions, would then be essential for communication and transmission of knowledge, ultimately leading to the transformation of old beliefs, that is habits, and to the formation of new ones. Habit is defined as a disposition, or motivation, to act in a certain manner under specific circumstances. Sign-function, then, is what determines the meaning of the sign based on the habits that the former generates, sustains and modifies. Meanings are to be verified in experience, but because a meaning as an analytic category is not reduced to psychological but encompasses the latter, it always already exceeds its own verifying instances; a meaning is therefore always further determinable.

Since semiosis is always a relational process comprising three categories, it effectively eliminates the dichotomy between subject and object and enriches the notion of interpretation "as more a matter of relation between signs than between signs and things." (Merrell 1995: 45). The dyadic behaviorist model gives way to the triadic, semiotic, model of human consciousness in "a sense of taking habit, or disposition to respond to a given kind of stimulus" (Peirce CP 5.440) even if the stimulus in question is barely liminal, as in the case of the category of mental images.

As a result of multiple interrelations, signs move from one to another, they grow and engender other signs because the triadic logic leads to signs always already becoming something else and something more, contributing – in the process of their growth – to human development and the evolution of consciousness. Platonic Forms, as some fixed indubitable truths, are only ideal limits. In their function as regulative ideas, they are signs, not essences, and therefore subject to evolution and growth. What fills a form, is a content – that is, meaning – the latter characterized by means of an experiential event rather than essence: "thoughts are events" (Peirce CP 5. 288).

By virtue of their meanings, "the ideas play a part in the real world" (Peirce MS 967. 1). The relationship between meaning and habit is one of reciprocal presupposition; habits have meanings, or rather *are* meanings (even if only structurally), but meanings may change depending on the formation of new habits; in turn, the new meanings eventually effect the habit-change, despite the fixed character of the latter. Precisely because of the fixed nature of habits, the abrupt change in meaning comes about by what Peirce identified as a cataclysm in the otherwise continuous evolutionary process.

Affirming the continuity of consciousness, Peirce stressed its temporal character. The cognitive, that is inferential, process of interpretation is a series of thought-signs, and the meaning of each thought becomes understood in each subsequent thought, creating a process of unlimited semiosis. No thought is ever instantaneous because it needs an inferential stretch for its own interpretation. Yet the immediacy of Firstness is always presented in an instant and, as Firstness, it is *had* prior to every mediative Thirdness, making inference seem to border on association and guessing.

Peirce, as long ago as 1868, stated that cognition exists only "in the relation of my states of mind at different instants.... In short, the Immediate (and therefore in itself unsusceptible of mediation – the Unanalyzable, the Inexplicable, the Unintellectual) runs in a continuous stream through our lives; it is the sum total of consciousness, whose mediation, which is the continuity of it, is brought about it by a real effective force behind consciousness" (Peirce 1955: 236-237) enabling the recursive process of *re-presentation* upon *presentation*.

The instance of Firstness embedded in Thirdness is manifest also in the Peircean propensity interpretation that assigns "an ontological status to the tendencies or propensities of the various possible outcomes of a singular chance event" (Shimony 1993/II: 237). Chance itself, in a pragmatic sense, functions in this respect as a principle (cf. Semetsky 2000a) because of any single variation's eventual function to possibly "bring about a change in every condition" (Peirce W4. 549 in Brent 1993: 175)[2].

As a sign of the uncertain character of the real *per se*, knowledge for Peirce is constituted by both observable and unobservable instances of general laws, that is, by both factual and counterfactual conditionals. For pragmatists, knowledge is greater than truth, and the continuity of inference, even if only in a probabilistic sense, defies the idea of some unknowable thing-in-itself, the latter being only hypothetical like any other First and is to be ultimately known – after "this is present to me" (Peirce CP 5. 289) – as a sign, or a Thirdness of Firstness. Signs reiterate, they become signs of signs, or re-presentations. As Peirce (CP 5. 138) stated, "the mode of being of a representamen [...a sign] is such that it is capable of repetition", that is, of creating sensible patterns. Yet, because every interpretant might be a precursor to a new meaning, different from the preceding one, the repetition is never the repetition of the same (cf. Semetsky 2000a). For Peirce, "our concepts ... literally 'participate' in the reality of what is conceived..." (Esposito 1980: 42) – implying holism and a sense of auto-referentiality between the inner and outer realities. Every sign is subject to interpretation by a series of subsequent thought-signs, and the whole triad enveloping the "the relation-of-the-sign-to-its-object becomes the object of the new sign" (Sheriff 1994: 37).

The value of knowledge is in its practical import, that is, the way we, humans, will act, think, and feel – in short, assign meaning to our own experience – as the pragmatic effect of the said knowledge. Negative exemplifications of the laws – that is, the instances of non-occurrence – are, according to Peirce, as effective as positive ones. Thus propositional knowledge must include generalizations independently of their having been already actualized in one's experience or taking place in the past, present or future. In Peirce's view the real is not in any way reduced to the actual, in fact "the will-be's, the actually-is's and the have-been's are not the sum of the reals. They only cover actuality. There are besides would be's and can be's that are real" (Peirce CP 8. 216), the would-be-ness constituting the realm of the virtual. The semiotically real world thus addresses possibilities; by the same token, the realist's view asserts the reality of potentialities not yet actualized.

Apart from the fact of being *a priori* unobservable, any potentiality also stays unobservable unless it turns into actuality as a consequence of some potential;

---

[2] Peirce's biographer, Joseph Brent (1993), notices the connection between Peirce's thinking and the science of chaos advanced by Ilya Prigogine who acknowledged a pioneering step made by Peirce toward understanding the physical world as characterized by sensitive dependency on initial conditions.

nonetheless for Peirce such "an admixture" (CP 1. 420) must be included in general law making. The real world therefore includes real potentials. The natural world becomes an object of interpretation, and the human cognition may be considered the necessary Thirdness in this relationship – for "man is nature's interpreter" (Peirce CP 7. 54) and both are embedded in the process of semiosis.

The dialogue, as an object for interpretation, is a relation between two signs. An utterer, as the producer of signs, is not reducible to a speaker. Nature, in its act of communication with human mind, is assigned the function of the quasi-utterer by Peirce, and mind, respectively, performs the function of a quasi-interpreter. The Thirdness of cognition thus governs Secondness, it determines the objects of knowledge as seconds because a Thirdness, performing a mediative function, creates or "brings information ... [it] determines the idea and gives it body" (Peirce CP 1. 537). The dialogue, being yet an empty form, becomes filled with meaning, that is substance, this process making the dialogue a sign of genuine communication.

The natural world, for Peirce, is *tychistic*, although at the level of seconds, among the brute facts of action and reaction, the physical laws conform to classical mechanics and space remains Euclidean. Considering however that the real world, contrary to the positivist thought, is not reducible to the category of Secondness, knowledge is always already fallible, and laws themselves are subject to evolution and change. "The idea of fallibilism objectified" (Peirce CP 1. 171) implies the diversity embedded in nature. What may bring about a change is chance itself, defying the absolute necessity of a cause. There are no axiomatic truths for Peirce, instead "chance, in the Aristotelian sense, mere absence of cause, has to be admitted as having some slight place in the universe" (W4: 546 in Brent 1993: 174) introducing a paradox of discontinuity and a possible change in direction within continuity itself.

Peirce asserted that all logical relations – hence, the process of semiosis – can be studied by means of being displayed in the form of existential graphs, or iconic representations; such diagrammatic thinking may yield solutions to the otherwise unsolvable logical problems, that is, also render a perplexing and problematic – in a social or psychological sense – situation solvable. Diagrammatic thinking takes place in the mind; it is an act of imagination both affecting and effecting human conduct in the world of matter produced as if by means of the formation of habits. All signs have a tendency "to affect certain others which stand to them in a peculiar relation of affectability" (Peirce CP 6. 104).

The meaning created by diagrammatic thinking is not actual but "altogether virtual ... [it is always contained] not in what is actually thought, but in what this thought may be connected with in representation" (Peirce CP 5. 289). Still(never mind its being virtual) it is maximally real because of the possibility of such a thinking being capable of producing real effects in terms of consequences, or "practical bearings" (Peirce CP 5. 402) in accord with the pragmatic maxim. Peirce considered consciousness a vague term and asserted that "if it is to mean

Thought it is more without us than within. It is we that are in it, rather than it in any of us" (CP 8. 256).

Everything is a sign: the whole universe, for Peirce, is perfused with signs; yet "nothing is a sign unless it is interpreted as a sign" (Peirce CP 2. 308). What seems to be a paradoxical statement is derived from the nature of the pragmatic method itself. The meaning and essence of every conception depends, in a pragmatic sense, on the way the latter is applied: it "lies in the application that is to be made of it" (Peirce CP 5. 532). In this respect, Jungian analysis – the subject-matter of the next section – not only involves the interpretation of signs as archetypal symbols constituting the realm of the unconscious, but borders on Peircean unlimited semiosis in terms of the archetypes' *"manifold meaning* [and] their almost limitless wealth of reference" (Jung CW 9i, 80).

On Jung's Analytical Psychology

The combination of words, *analytical psychology*, may seem to be a contradiction in terms if not for remembering that Peirce's semiotics blurs the boundaries between logic and psychology or, we may add, blends them both into an area of interest for the cognitive science; Peirce, in fact, has introduced the distinction between *token* and *type*. Signs are not merely tokens of the actual semiotic process. As types, they belong to the potential field of being used in some lawful manner, and pragmatically capable of producing an effect as a result of their usage. In this respect, archetypes of the collective unconscious and the acausal connective principle, or synchronicity, posited by Jung (CW 8) in 1952, are not solely mystical entities.

What is required, is a change in conception. An acausal connection seems to be an illogical statement; for Peirce, however, a paradoxical, that is, "a self-contradictory proposition is not meaningless; it means too much" (CP 2. 352). Jung rejected the solely dyadic logic and, similar to Peirce, asserted that "psyche and matter are two different aspects of one and the same thing" (Jung CW 8, 418). As if anticipating the post-Cartesian philosophies, Jung did not draw a line of great divide between the products of imagination and those of intellect: both affect thinking, and all thinking aims at the creation of meanings. This section attempts, by employing the three Peircean ontological categories, to unpack the complexities of Jungian conceptualizations.

Briefly, in its practical sense, Jungian analysis incorporates "the paradigm of an *active, interventionist therapist*" (Samuels 1985: 197) who facilitates an analytic session by means of *interpreting* images that may appear as unconscious material in the analysand's dreams, or art forms like pictures and drawings, or in the course of an active imagination during sessions. At the level of theory, the unconscious, for Jung, is not reduced to Freudian repression in a personal sense, but is specified as lacking meaning, that is, as yet – prior to the Thirdness of mediation – being out of the conscious awareness. Still, considered as a sign, or relational entity, the unconscious *per se* belongs to the Firsts and does function as a powerful and "real effective force behind consciousness", as we said earlier citing Peirce.

The unconscious is collective, that is, it is based on the experiential heritage and history of all humankind, and its content is determined by the activity of archetypal dynamic patterns, indeed "habits-taking" (Peirce CP 1.409), manifesting as universal motifs in human behavior. Habits, for Peirce, are dispositions to act in a certain way under specific circumstances "and when actuated by a given motive" (CP 5. 480). As for unconscious archetypes, they were conceptualized by Jung as being "a real force charged with specific energy" (1963: 352). Acknowledging their potential effect on human conduct, Jung also defined archetypes as "system[s] of readiness for action" (CW 9, 199).

A sign, "in order to fulfil its office, to actualize its potency, must be compelled by its object" (Peirce CP 5. 554), therefore it strives to appear in a mode of Thirdness and become available to integration into consciousness. We remember Peirce's assertion that a sign is in fact a sign if - and only if - it is interpreted. An act of imagination is potentially transformative, according to Peirce, in its function as deliberation for the purpose to generate a meaning for a habit. For Jung too, the archetypal images are "endowed with a generative power; ... [the image] is psychically compelling" (Samuels, Shorter & Plaut 1986: 73).

The habit that manifests itself in a particular way of human cognition, as well as conduct, and includes one's emotions and perceptions, may become identified in the course of the analytical relationship as embedded in some actual problematic situation. The situation is problematic, that is, it involves tension and conflict, because it encounters the otherness, or Secondness of "reaction against my will" (Peirce CP 8. 144) due to the intervention, sometimes beyond one's awareness of this action, of the brute facts of human experiences. The purpose of analysis consists of *individuation* which is seen as a process of integration of conscious and unconscious aspects of one's self for the "achievement of a greater personality" (Jung CW 7, 136).

Integration, as the production of meanings, leads to potential change in one's habitual ways of thinking, feeling and behaving as eventual effects of the analytic process, the latter based on archetypal imagery embedded in the collective unconscious. Thus Jungian analytical psychology, both theoretically and practically, may be considered a pragmatic method (cf. Noddings 1993: 105) that accords with Peirce's maxim: "Consider what effects, that might conceivably have practical bearings, we conceive the object of our conception to have. Then our conception of these effects is the whole of our conception of the object" (CP 5. 402).

New information, derived from the unconscious material as the effect of interpretation, not only determines the idea conceptually but also "gives it body" (Peirce CP 1. 537) in the world of action. The archetypal images are thus semiotic by virtue of their being carriers of information (see Semetsky 1998, 1999b) embedded in the collective unconscious: the unconscious is capable of spontaneously producing images "irrespective of wishes and fears of the conscious mind" (Jung CW 11, 745). The archetypes are postulated by Jung to serve as a psychological, or, better, psychic – hence, *depth* psychology – ground for habits. Never mind their status as ideas, or rather because they are indeed regulative and generative ideas, the archetypes as *symbols* of transformation are effective in the *physical* world: we remember Peirce's having said that "the ideas do play a part in the real world" (MS 967. 1).

Mental images are not pure icons, they become enriched with indexicality; they perform a definite sign-function and point to some antecedent event contributing to their appearing in the unconscious. Thus they *indicate* Seconds of actions and reactions, rather than just being Firsts of the as yet disembodied mind. Jung used the word *symptom* (cf. Sebeok 1991) within clinical discourse. However symptoms do not serve merely a diagnostic purpose. The collective unconscious encompasses future possibilities, and "[a] purposively interpreted [image], seems like a *symbol*, seeking to characterize a definite goal with the help of the material at hand, or trace out a line of future psychological development" (Jung CW 6. 720), that is to perform a prospective, prognostic function.

In this respect, Peircean might-be-ness and would-be-ness, that is his altogether virtual Real, seem to be isomorphic with the realm of the collective unconscious, the latter defined not only as the repository of human past, inheritance, dispositions, but also future developments. Jung's position appears to affirm the concept of final causation[3] in his saying that "the archetype determines the nature of the configurational process and the course it will follow, with seeming foreknowledge, or as if it were already in a possession of the goal" (Jung CW 8, 411). The archetype's function is that of a Peircean "general idea …[which] is already determinative of acts in the future to an extent to which it is not now conscious" (Peirce CP 6. 156).

The synthesis of time inscribed in the collective unconscious as the universal memory pool accords with Peirce's semiosis acting within a shared layer of human experiences that includes dimensions of past, present and future: "A man denotes whatever is the object of his attention at the moment; he connotes whatever he knows or feels of this object, and is the incarnation of this form …; his interpretant is the future memory of this cognition, his future self, or another person he addresses, or a sentence he writes, or a child he gets" (Peirce CP 7. 591). The Interpretant as a Third, therefore, has an anticipatory, albeit vague, due to the presence of Firstness in itself, flavor.

The Thirdness of interpretation in its mediation performs the amplifying function, constituting the basis of the Jungian *synthetic* method which implies *emergence* – that is a leap to a new meaning (cf. Peirce) – as carrying the utmost significance. Synthetic method thus reflects the future-oriented path to knowledge, the memory of the future or what Jung called a prospective function of the unconscious, and indeed amplifies traditional psychoanalysis which was

---

[3] In Semetsky 2000d, I address in detail the concept of self-cause as represented by one of the major cards in the deck, "The Magician".

considered by Jung as reductive because of its sole orientation to the past marked by a single signified.

For Jung, as for Peirce, "psychological fact ... as a living phenomenon, ... is always indissolubly bound up with the continuity of the vital process, so that it is not only something evolved but also continually evolving and creative" (Jung CW 6: 717). Moreover, Jung's defining the collective unconscious as the *objective* psyche outside actual personal experience and his notion of archetypes that may appear as mental representations of an object – even if the latter is, as we said earlier, "a pure fiction" (Peirce CP 4. 351) – describe in a way "the Reality which by some means contrives to determine the Sign to its Representation" (Peirce CP 4. 536).

The reality *contriving* to determine the sign to its representation is, for Jung, the psychic reality: as a sign, the very depth of the psyche creates a *relation* between the worlds of mind and matter. Mental images as icons are *immediate* objects in their Firstness – we remember Peirce's saying "this is present to me" (CP 5. 289) – but the archetypes to which they refer, seem to accord with the Peircean definition of the *dynamical* object "which ... the Sign ... can only *indicate* and leave to the interpreter to find out by *collateral experience"* (CP 8. 314).

For Jung, archetypes are general tendencies and subsist, rather than exist, *in potentia* only. As skeletal concepts, their significance is not exhausted by Platonic Ideas: as Firsts, they are only "forms without content, representing merely the possibility of a certain type of perception and action" (Jung in Spinks 1991: 448). But the vague and unconscious forms are to be filled with informational content embedded within real, flesh-and-blood, human experiences in the phenomenal world. Situated in the midst of the Seconds, within real human actions and reactions, they need thought and interpretation as Thirds so as to acquire meaning by virtue of being "altered by becoming conscious and by being perceived" (Jung in Pauli 1994: 159). The plurality of evolving meanings find their expression in the symbols of transformation comprising, in a Peircean sense, a series of thought-signs and sign-events. A symbol, for Jung, "points beyond itself to a meaning that is ... still beyond our grasp, and cannot be adequately expressed in the familiar words of our language" (Jung in Noth 1995: 119) but needs a medium – Thirdness – for its expression.

The relationship between the collective unconscious and individual consciousness is of utmost importance for Jung. Signs are "always grounded in the unconscious archetype, but their manifest forms are moulded by the ideas acquired by the conscious mind. The archetypes [as] ... structural elements of the psyche ... possess a certain autonomy and specific energy which enables them to attract, out of the conscious mind, those contents which are better suited to themselves" (Jung CW 5: 232) – that is, as Peirce would have said, potentially "connected with in representation" (CP 5. 285). The attraction is a quality of affect, the latter – in its relation to Firstness – is indeed independent, that is, autonomous.

In order to explain the significance of contingent events as meaningful coincidences, Jung postulated the so called synchronicity principle, that is the absence of a direct (or local, in the language of contemporary physics) cause-effect connection. Recall Peirce's asserting that the absence of cause had to be admitted as playing a part in the universe: although Jung never overestimated the role of pure chance, his was indeed an acausal connecting principle attempting to overcome the chance/cause dualism and to explain the occurrence of coincidental events as having value and meaning.

The principle of synchronicity was developed by Jung in collaboration with Wolfgang Pauli who has taken the idea seriously and elaborated on it in detail. Synchronicity addresses the problematic of meaningful patterns generated both in nature and in human experience, linking the concept of the unconscious to the notion of " 'field' in physics ... [and extending] the old narrow idea of 'causality' ... to a more general form of 'connections' in nature" (Pauli 1994: 164). Pauli envisaged the development of theories of the unconscious as overgrowing their solely therapeutic applications by being eventually assimilated into natural sciences "as applied to vital phenomena" (1994: 164).

Referring to various phenomena that may appear random and senseless if not for their meaningful synchronistic significance, Jung has stated that "it also seems as if the set of pictures in the Tarot cards were distantly descended from the archetypes of transformation" (CW 9, 81). This brief note has subsequently inspired a substantial body of work produced by contemporary post-Jungians (see Semetsky 1994, 1998). Andrew Samuels, for example, mentions "systems such as that of the *I Ching*, Tarot and astrology" (1985: 123) as probable, even if questionable, resources in analysis and quotes Jung's writing in 1945: "I found the *I Ching* very interesting. ... I have not used it for more than two years now, feeling that one must learn ... or try to discover (as when one is learning to swim) whether the water will carry one. (quoted in Jaffe 1979)" (Samuels 1985: 123).

Jung's biographer Laurens van der Post, in his introduction to "Jung and Tarot: an Archetypal Journey" by Sallie Nichols (1980), notices the contribution to analytical psychology made by "Nichols, in her profound investigation of Tarot, and her illuminated exegesis of its pattern as an authentic attempt at enlargement of possibilities of human perceptions" (1980: xv). Irene Gad (1994) has connected Tarot cards with the process of individuation and considered their archetypal images "to be ... trigger symbols, appearing and disappearing thoughout history in times of transition and need" (1994: xxxiv). The following section not only grounds Tarot in Jungian archetypal symbolism, but also describes its structure in terms of a self-organizing system of signs and suggests that its dynamics accord with the principles of second-order cybernetics.

On the Structure and Dynamics of Tarot.

Peirce has stated that "the most perfect of signs are those in which the iconic, indicative and symbolic characters are blended as equally as possible" (CP 4.

448). Such an optimal combination is displayed in the sign-system of Tarot cards. Pictures are by definition iconic, each icon being rich with the elements of symbolic imagery, and the cards' layout is indexical by virtue of its pointing to the archetypal field of the collective unconscious.

We remember that archetypes, for Jung, may appear as images in one's psyche despite the archetypes themselves existing only in their potential virtual state and being unable by themselves "to be capable of reaching consciousness" (Jung CW 8. 417). We also remember Peirce's asserting that for a sign to be a representamen of an object, the latter does not have to solely have an actual physical existence but may be a thought or a mental image even if altogether fictive. The pilot-study (Semetsky 1994) has provided substantial empirical data warranting those assertions.

Each Tarot image positioned in a layout may be considered "an Icon of a peculiar kind" (Peirce CP 2. 248). Because of its a-priori indexicality in its relation to the archetypes of the collective unconscious, the Tarot layout performs the function of "rendering literally visible before one's very eyes the operation of thinking *in actu*" (Peirce CP 4. 571). Functioning in its iconic-indexical form in the mode of Peircean existential graphs, a layout therefore asserts "the epistemological thrust" (Spinks 1991: 446) of the latter pointing towards the epistemic implications of the Tarot sign-system. As enabling some access to the yet unconscious, tacit, "knowledge", the interpretation of Tarot images contributes to the "construction" of the former via its mediation by means of pictorial language so that this "knowledge" becomes available to consciousness.

In this manner, Peircean symbolic logic, by means of being embodied in its own visual notation, does contribute to reasoning and the creation of meanings implicit in the layout through the thirdness of interpretation. In its totality, therefore, Tarot is semiotic *par excellence,* and in its traditional, albeit limited, sense as cartomancy, it has been defined "as a branch of divination based upon the symbolic meaning attached to individual Tarot cards or modern decks, interpreted according to the subject or purpose of a reading and modified by their position and relation to each other from their specific location in a formal 'layout' or 'spread'" (in Sebeok 1994/I: 99).

The reference to *semiosis* as a dynamic *action* of signs does not however appear in this definition. We remember that the presence of Peircean quasi-interpreter and quasi-utterer – even if, we add, in non-verbal, pictorial language – is necessary for the dialogic communication between mind and nature. The symbolic lexicon (see Semetsky 2000b) of pictures comprising the layout becomes a representation of a necessary Thirdness – indeed, and following Jung, beyond the aforementioned "familiar words of our language" – which is paramount for the production of meanings.

As signs, that is, having a triadic relational structure, archetypes do demand "a naturalistic interpretation" (Laszlo 1995: 135). But not merely from the perspective of semiotics. Asserting their naturalistic status, Laszlo approached the archetypes from a systems science point of view and respectively argued that they, as well as "the collective unconscious that frames them, are not just 'in the mind': they are in nature" (Laszlo 1995: 135), making Jungian psychic reality conceptually analogous to "the collective, space-and-time-dimensionless"[4] (1995: 136) field in nature.

The Thirdness of interpretation, as we said earlier, creates or "brings information ... [it] determines the idea and gives it body" (Peirce CP 1. 537). The idea of Tarot as *embodied* mind has been addressed by the general systems theorist Erich Jantsch (1975), who has included Tarot in his systematic overview of approaches and techniques of the so called inner way. Jantsch placed archetypes and Tarot at the mythological level among genealogical approaches and, emphasizing the continuous self-organization of systems through self-realizing and self-balancing processes, has noted that "Tarot cards ... may be seen as embodying ...[and] mapping out the field of potential human response" (Jantsch 1975: 163; see also Semetsky 1999b).

In its capacity as a semiotic system, however, Tarot can be moved up to a level identified by Jantsch as evolutionary, and at which he acknowledged the human potential of being capable of "tuning in" (1975: 150) to the aforementioned field in nature, thus overcoming the limitations of three-dimensional space or phenomenal time. As for the complicated task of tuning in to such a field, Jantsch anticipated, among other things, a dynamic "communication mechanism, which is at work across the ... levels of perception, so that, for example, 'insight' from the evolutionary level may be received in some other form at the mythological level, e.g., in the form of intuition, or dreams, or general vibrations felt as quality" (Jantsch 1975: 149). Such a communication mechanism, a semiosis in nature, is embedded in a Tarot layout comprising a sequence of pictorial cards that are to be "read", that is narrated and interpreted.

The transformational pragmatics of Tarot, that is the potential effect of each reading, is provided by means of the intervention of Thirdness of interpretation as conducted by a "reader". The reader's interpretive strategies[5] are based on the principle of polysemy: there is no fixed meaning attached to a particular card, rather the meanings are contextualized depending on a particular situation, as well as derived from a position occupied by this or that card in a layout. While some positions in a spread traditionally correspond to specific propositional attitudes and describe such common semantic categories as beliefs, fears and hopes, the archetypal content of each representation, that is the information inferred from the card's imagery, will vary as a function of its place, or *topos*.

For Jung, there are as many archetypes as there are typical situations in life. We remember Peirce's referring to a trace of the future self, or another person, or a child, or a sentence one writes, that is to semiotic information as represented in

---

[4] Or rather multidimensional? Perhaps synchronistic phenomena may be partially explained by the hyperspace theory.

[5] Here I describe the appoach based solely on my personal experience in my capacity as a reader. This paper does not make any claims towards generalizations regarding specific reading techniques as pertain to other Tarot readers. See Semetsky 2000b for some relevant research.

signs. Sure enough, although the number of cards in a deck is finite, their permutations and combinations tend toward infinity, reflecting the richness and unpredictability of human experiences and associated affective states. What may be "predicted", though, is the *tendency* for event to occur or a singular state of the system which embodies the corresponding informational content. Interpretation contributes to *trans*-formation of *in*-formation – through indexical connection – from the unconscious into the conscious, implying a possibility of not only habits-taking but also habits-breaking! This transformation, or habit-change, would be practically improbable if not for the future acting upon the present, being pulled into the present by archetypal forces that play the role of "inward [or] *potential* actions ... which somehow influence the formation of habits." (Peirce CP 6. 286).

Jantsch (1980), from his systems-theoretical perspective, acknowledged this somewhat backward causation as a self-organized system's feature of anticipation: the system's present state contains "not only the experience of past evolution, but also the experience of anticipated future [that] vibrates in the present" (Jantsch 1980: 232). Some positions in a typical spread correspond to what Jantsch described as "the fine-structure of time" (1980: 232), a singular layout combining in itself all three aspects of time. The layout reflects on the possibility of anticipating the future by enabling the feeling of peculiar "gazing" into the possible future which may be described as "the options in further evolution" (Jatsch 1980: 232.) in the system's dynamics.

As Thirdness, the Tarot spread itself mediates between one's consciousness and the collective unconscious, thus serving as a logical interpretant *of* the latter, as well as a dynamical interpretant *for* the mind of the subject of a reading, that is a person for whom a reader indeed "reads", that is interprets, the informational contents of this person's mental representations contained in archetypal Tarot images. So in the material world the *structure* of a layout seems to appear as if from nowhere, by synchronicity, but in fact appears out of the *process* implicit in Jungian psychic reality. This process, as the Peircean category of Thirdness, indeed governs Secondness and creates or brings information. Jungian synchronicity seems just another name for distant connections between events in nature, suggesting that "an acausal connection may manifest itself in the form of non-local correlations that appear to lie outside the normal confines of space and time" (Peat 1992: 199).

In other words, the *static structure* of the layout may be considered a projection, in the sense of projective geometry, or a snapshot[6] of a *dynamic process* as the very action of signs. This conceptualization accords nicely with Tarot's psychological function as a kind of projective technique (see Semetsky 1994) or a psychological tool that not only parallels but even exceeds the Rorschach method used in clinical practice for the purpose of assessment and

---

[6] See Semetsky 1999a for the relations of movement and rest in Tarot semiotics from the perspective of Gilles Deleuze's philosophy of transcendental empiricism.

testing. By definition, the projective method is viewed as a structured interview or a dialogue, that is, an open and flexible arena for studying interpersonal and intrapsychic transactions. As paradigmatic Thirdness, Tarot cards, in their projection into layout, display the triadic quality of representation, relationality and mediation, thus constituting what Peirce called "a portraiture of Thought" (Peirce CP 4. 11).

While *reading* is a conventional term for interpreting the Tarot spread, the meaning of it seems to come close to what in contemporary cognitive science (Von Eckardt 1996) has been called a *theory of content determination* for the human mental representations system, especially with regard to their psychological grounding, that is habits, from a Peircean perspective. Each Tarot image, by definition, is a sign. As such, and in the almost animate manner of "living signs" (Merrell 1999: 453), it "*endeavors* to represent, in part at least, an Object which is therefore in a sense the cause, or *determinant* of the sign" (Peirce CP 6. 347; italics mine, IS) hence ensuring the appearance, by virtue of the very action of signs, of this card in the layout.

Conforming to Peirce's theory of signs in action, or semiosis, the following procedure takes place during a reading: first, it is assumed that one's mental representations are semiotically encoded in the laid-out cards, comprising their informational content; second, the cards' imagery is connected, for the purpose of interpretation, with the background of specific archetypal patterns, constituting one's unique and complex world of experience as it pertains to the present context and situation; and, third, the interpretation creates a field of meanings for the mental representations as a form of narrative knowledge.

Jung maintained that in analysis "every interpretation necessarily remains 'as-if'"(Jung CW 2, 265). Indeterminacy abound, "certain fundamental meanings ... can only be grasped approximately" (Jung CW 8, 417) in agreement with Peirce's asserting any prediction as being of general and incomplete character. Ultimately however, if "certain sorts of ink spots... have certain effects on the conduct, mental and bodily, of the interpreter" (Peirce CP 4.431 quoted in Von Eckardt 1996: 151), then it is quite logical to assume that eventually interpretation will lead to habit-change according to some lawful relationship. Thus Tarot acquires not only syntactic and semantic dimensions, but pragmatic as well, in the sense of rendering the flow of information meaningful by means of its effectiveness with respect to future consequences.

Peirce asserted the possibility of transformation not only at the mental level but at the level of actions: habit-change means "a modification of a person's tendencies toward action" (CP 5. 476), such a modification being the ultimate purpose of the reading process. Habits, however, are resilient – they wouldn't be habits otherwise – and their function is similar to the action of archetypes that, according to Jung, can sometimes possess the psyche in a guise of an individual or collective *Shadow*[7] – the latter corresponding, incidentally, to card number XV, "The Devil".

Von Eckardt's insight that "we do not *use* our propositional attitudes. Rather, they themselves involve a 'use' of, or an attitude toward, a content" (1996: 165), seems to imply the archetypal, that is bordering on possessive and forceful, nature of mental states. Archetypes can be "the ruling powers" (Jung CW 7, 151). The encounter with one's own powerful *shadow* constitutes a fundamental part of Jungian analysis. As for Tarot, a single deck consists of seventy eight cards, including twenty two of the so-called major arcana, of which the Shadow archetype is just one. The combination of many cards in an individual layout enriches the analytic sessions – readings – with the greater amount of available information derived from the unconscious which comprises the complex interplay of often conflicting emotions and behaviors.

The Tarot system functions as a dynamical interpretant by virtue of its being a sign that stands for one's real emotional, behavioral and cognitive patterns expressed via the symbolism of the cards (Semetsky 1994, 1998). Jung commented that "our brains might be the place of transformation, where the relatively infinite tensions or intensities of the psyche are tuned down to perceptible frequencies and extensions" (Jung in Laszlo 1995: 135) so as to enable the reading. Due to the mediating function of interpretation, the latent, unconscious, contents of the mind are rendered conscious, and the signs which are brought to the level of awareness, that is, intensified and amplified up to the point of their possible integration into consciousness, are capable of creating a momentous feedback in the psychodynamic processes of the subject of a reading.

The present psychic structure – or a subject's current level of self-knowledge – tends to some instability threshold. Indeed, the aforementioned "integration is not continuous but rather marked by the kind of discontinuities and phase transitions associated with complexity theory, as formulated, for example by Thelen and Smith (1994)" (Muller 2000: 59; cf. Semetsky 1998); these dynamics capable of producing "a change in the subject's mental life which, in turn, changes his or her disposition to act... in ways dependent *on the content of representation*" (Von Eckardt 1996: 283-284). The irreversible change takes place in the actual physical world asserting the objective, as stated by Jung, reality of archetypes and also warranting Peirce's pragmatic maxim as the production of real, not just metaphorical, effects.

"The habit alone" (Peirce CP 5.491), as a logical interpretant, is capable of abruptly interrupting the semiotic regress, effecting its own transformation by the operational "closure of the process, ... a closure which itself opens possibilities" (Colapietro 2000: 145). Indeed, the *explication* of the information *implicated* in the collective unconscious tends to its *complication* with regard to an individual consciousness, thus attributing to semiosis the potential function of contributing to human learning – especially by means of—"the deliberately formed, self-analyzing habit" (Peirce CP 5. 491) – the latter not only being embodied in the Tarot spread, but also ultimately veritable in one's experience.

The human psyche, functioning as "unextended intensity" (Jung in Laszlo 1995: 135), is typically marked by tensions or bifurcations which signify "a fundamental characteristic in the behavior of complex systems when exposed to high constraint and stress" (Laszlo 1991: 4). During readings, a Tarot layout may indicate the presence of a highly unstable situation or a state of mind, although the mind itself, at the conscious level, may be quite unaware of its own situation but still *feel* the latter's emotional impact or, within a clinical discourse, be in a certain affective state. We remember Peirce's saying that the brute facts of life may intervene quite "against [one's] will" (CP 8. 144). The outcomes of such a tension imposed on a system will vary: similar to the bifurcations classified according to their degree of manifestation, as well as the dynamic regime in which a system will potentially settle, various major cards are signs of either subtle (e.g. "The Wheel of Fortune"), catastrophic (e.g. "The Death"), or even explosive (e.g. "The Tower"; see Semetsky 2000c) bifurcations.

The semiotic bridge, established by means of a synchronistic connection between the collective unconscious and an individual mind, enables insight into the meaning of a current situation, indeed "makes sense" out of it. "An acausal parallelism" (Jung 1963: 374) of synchronicity would perhaps, in Peircean classification, be classified as a precognitive quali-signification, that is the qualitative immediacy of experience. The immediate Firstness – a sort of premodern natural attraction – was, together with the Thirdness of mediation, left out as insignificant by modern science and substituted by the dualistic sin-signification and instrumental rationality based on conventional logic of excluded middle. That is why the readings enable, as we said earlier, re-presentation upon presentation: a layout prior to the Thirdness of interpretation is the presentation of unconscious – by virtue of semiosis – and *not* yet representation of one's conscious mental states.

The actualization – via "magnitude of thirdness" (Deely 1990: 102) – of many potentialities, the Firsts, of the unconscious, is taking place due to the subjective, bottom-up, "intervention of the mind" (Shimony 1993/II: 319) of the interpreter into a signifying chain of semiosis. Yet this very intervention may be considered objective in the sense of itself being implemented by a choice of a global, top-down, character, analogous perhaps to the semiotic functioning of "a *relationis transcendetalis*" (Spinks 1991: 444). A choice of this kind may be accounted for by means of what Shimony, addressing "the status of mentality in nature" (Shimony [Penrose, Shimony, Cartwright, Hawking] 1997: 144), dubbed a (hypothetical) super-selection rule in nature that enables the very "transition between consciousness and unconsciousness ... not ...as a change of ontological status, but as a change of state" (1997: 150).

The absence of an ontological dualism also presupposes the non-local "contact with some sort of Platonic world" (Penrose [Penrose et al.] 1997: 125) of archetypes[8]. The relationship, as posited by Penrose, between three worlds,

---

[8] The Platonic world as addressed here by Penrose is properly philosophical and is not reduced to mathematical truths but involves the good and the beautiful, that is "non-computable elements – ...judgement, ...compassion, morality,..." (1997: 125). These are all archetypal meanings reflected in the images of tarots. Card # XX, for example, is called "Judgement".-

namely the physical world, the world of ideas, and the world of mind, has been considered a mystery and heavily debated by the computational strand of contemporary cognitive science as "gaps in Penrose's toilings" (Grush & Churchland 1995). The core of Penrose's argument is that the physical world may be considered a *projection* of the Platonic world of ideas, and the world of mind arises from part of the physical world, thus enabling one in this process to grasp and, respectively, understand some part of the Platonic world.

The triadic relationship implicit in Penrose's diagram is indeed semiotic. The Platonic world, however, is not the world of perfect geometrical solids. The archetypal *eideas*, according to contemporary post-Jungians, are considered to be both structuring patterns of the psyche and dynamic units of information, and seem to be modeled on strange, or chaotic, attractors (see Van Eenwyk 1997) and Mandelbrot's fractals rather than on Platonic ideal forms. Jung himself is viewed as a systems-theorist, and "a systemic ... view implies that ... inner and outer, ... interpersonal and intrapsychic can be seen to be just that seamless field of references" (Samuels 1985: 266) which we may call semiosis.

The transformational pragmatics of Tarot is effected because of the included middle of Peircean Thirdness, but the interpretation itself is triggered by abduction that is always already present as Firstness and without which no hypothesis-making is possible. This Firstness in Thirdness is being "tested" and deliberated upon during the reading among the continuous interplay of all three forms of inference. The path to knowledge is not confined to pure reason, and what makes Tarot reading efficient is perceptual judgement being a sort of "mediated immediacy" (Peirce CP 5. 181), which as a limiting case of abductive inference – an educated guess – indeed triggers the reading. The process of reading accords with Peircean diagrammatic reasoning when a reader, passing from one picture in the layout to the other, from one icon to yet another, "from one diagram to the other ... will be supposed to see something ... that is of a general nature" (Peirce CP 5. 148).

The abductive guess as a matter of a First borders on intuition; an intuitive knowledge traditionally being a synonym for immediate knowledge. Intuition for Peirce does mean cognition, however the latter is determined by the object outside one's personal *cogito*. While the conclusion of an argument is clearly determined by other cognition, intuition has been considered to be the initial perception of an object. Yet for Peirce, there is no immediate, that is unmediated, knowledge: all cognition is mediated by signs in a process of semiotic inquiry. Thus perception differs not in kind but only in degree from other forms of human knowledge.

Abduction seems to function instantaneously, not because there is no temporal interval of inference, but because the mind is unaware of when it begins or ends. Peirce, describing the structure of perceptual abduction, noted that "the first premise is not actually thought, though it is in the mind habitually. This, of itself would not make the inference unconscious. But it is so because it is not recognized as an inference; the conclusion is accepted without our knowing how" (CP 8. 64-65). Intuition, albeit achieving an intellectual knowledge, the *nous* of the ancients, is indeed not *of* something but *is* something; as an epistemic pragmatic method, it is a process of *knowing* rather than knowledge.

Developing one's own intuition is a challenge for a reader (see Semetsky 1994, 1999a), and the information from the collective unconscious, outside the *cogito*, widens the boundaries of an individual's consciousness, contributing to the organization of the latter at a higher level of complexity. In Tarot *in-tuition* functions in accordance with its literal meaning, that is learning from within, from the very depth of the psyche, affirming its place in the semiosis as "communication ...across the ...levels of perceptions", as we said earlier citing Jantsch (1975: 145). Access to knowledge then, "and this is a crucial point, is available *within* ourselves" (Jantsch 1975: 146) as much as without, making the aforementioned *relationis transcendetalis* in fact immanent in perception!

The latent, unconscious, contents of the mind become available to cognition and are therefore rendered conscious because of the structural coupling effected by interpretation. So a reading performs a strictly auto-referential function: "it addresses somebody, that is creates in the mind of the person an equivalent sign or perhaps a more developed sign" (Peirce CP 2. 228), notwithstanding that "the first sign" (ibid.) is still the same — yet different, because one is not conscious of oneself — "somebody". The layout is created — indeed, *laid out* — as the ultimate "interpretant of the first sign" (ibid.). The interpretant stands for its object "in reference to a sort of idea" (ibid.) as the mind's archetypal ground. Tarot then performs a double function of being both a sign-object for the signs which act in, or in-*habit,* the collective unconscious, and a sign-interpretant contributing to one's *habit*-change.

We do remember Jung's pointing to amplification in analysis. Another function operative in psychological process was, as Jung noted, compensation, that is a tendency of the unconscious to maintain balance and stay in a homeostasis with the conscious mind for the purpose of self-regulation. In other words, and in terms of information theory, there is a natural presence of a *negative* feedback necessary for adaptation. In a somewhat different sense Peirce used the terms *ampliative* and *explicative* to distinguish between those forms of reasoning that aim at increasing knowledge and, by contrast, to make hidden or implicit knowledge explicit, to make manifest what is latent.

In this respect, Tarot functions in the two-fold manner of second-order cybernetics: both as an amplifier by rendering the subtle aspects of one's psyche vivid and substantial, and as a *positive* feedback that directs the amplified information back into the system, thus equipping it with an increase in information by having made the latent unconscious contents manifest and rendering them meaningful. Indeed, what is implicated in the mind is not only explicated but becomes complicated as well. The double contingency (cf. Luhmann 1995) embedded in Tarot self-organizing dynamics leads to a new level in a system's organization; the surpluses of information — because the collective unconscious by its very definition always already includes the personal unconscious — lead to learning and an increase in a system's complexity.

The double-folding is a feature of a non-linear evolutionary process. The latter, according to Luhmann's "systems-theoretical view-point ..., is a circular process that constitutes itself in reality (and not in nothingness!) ... Every system that participates in interpenetration realizes the other within itself as the other's difference between system and environment, without destroying its own system/environment difference" (Luhmann 1995: 216). This means that only by participating in multiple interactions and transactions, the number of which tends toward infinity can an individuation, or self-realization, constituting the aim of Jungian analysis, be achieved.

Individuation, as a never-ending process toward the maximally integrated personality, was used by Jung in the same sense as the Peircean term individual: "the identity of a man consists in the *consistency* of what it does and think" (Peirce 1995: 250) as embedded in the collective process. Personal wholeness is an ideal limit approximated by actualities in the hecceities of experience. Still, because "consistency belongs to every sign, ... the man-sign acquires information and comes to mean more that he did before" (Peirce 1955: 249). Thus, an active participation in semiosis by entering its *consistency* through its own symbolic representation in tarot signs may, from the perspective of evolutionary science, contribute to human growth in a continuous process of becoming other and hopefully "more fully developed" (Peirce CP 5. 594) signs among signs.

Conclusion

Pragmatic philosophy is consequentialist, which means that actions are evaluated according to the consequences they produce. It this respect the ethical question arises of how to treat the information that becomes available as a result of readings and implies, by virtue of its being a motivational force behind the transformation of habits, a possibility of producing new modes of action in the social world.

The interpenetration of epistemology and psychology leads to both acquiring ethical connotations. The ethics of care (Noddings 1984) becomes a must. Care theorists recognize not the abstract universals of moral philosophy but only those arising from concrete human conditions: "the commonalities of birth, death, physical and emotional needs, and the longing to be cared for. This last – whether it is manifested as a need for love, physical care, respect or mere recognition – is the fundamental starting point for the ethics of care" (Noddings 1998: 188). Those common human feelings and desires are inscribed in tarots, each card being a reflection of the archetypal force subsisting in the collective unconscious. Psychic residues are formed by recurrent experiences and are laid down in archetypal structures, but – we repeat, as "system(s) of readiness for action" (Jung CW 9. 199) – those structures themselves can in an auto-referential manner "exert an influence on experience, tending to organize it" (Samuels et al. 1986: 24), thus effecting changes and creating new possibilities in the experiential world.

This caring attitude based on relational ethics will also have to, by definition, respect the semiotic "anomaly" of tarots (see Semetsky 2000b). The process of semiosis seems to allow for "the emergence, in human brains, of holistic structures that can mirror, simultaneously, both the *structural forms* and *functional effects* of human thoughts" (Stapp 1993: 178), affirming the Jungian definition of archetypes as dynamical, or process-structures, of the psyche. The interplay of signs thus in a very significant way acts "both to veil the form of fundamental reality and to unveil the form of empirical reality. However if causal anomalies actually do appear then the veil has apparently been pushed aside; we have been offered a glimpse of the deeper reality" (Stapp 1993: 181). This deeper reality is the Peircean, semiotic, reality that appears to have found its own means of communication in the pictorial language of Tarot signs.

References

Brent J. (1993). *Charles Sanders Peirce, A Life.* Bloomington and Indianapolis: Indiana University Press

Colapietro, V. (2000). "Further Consequences of a Singular Capacity". In *Peirce, Semiotics, and Psychoanalysis.* John Muller and Joseph Brent, (Eds.), 136-158. Maryland: The John Hopkins University Press.

Deely, J. (1990). *Basics of Semiotics.* Bloomington & Indianapolis: Indiana University Press.

Downing, D. (1995). *Dictionary of Mathematical Therms.*(Second Edition). New York: Barron's Educational Series, Inc.

Esposito, J. (1980). *Evolutionary Metaphysics: The Development of Peirce's Theory of Categories.* Athens: Ohio University Press.

Griffin, David R., (Ed.). (1993). *Founders of Constructive Postmodern Philosoph: Peirce, James, Bergson, Whitehead, and Hartshorne.* Albany: State University of New York Press.

Gad, I. (1994). *Tarot and Individuation.* York Beach, ME: Nicholas-Hays, Inc.

Grush, R., and Patricia S.Churchland (1995). "Gaps in Penrose's Toilings", *Journal of Consciousness Studies,*2, No.1., 10-29.

Jantsch, E. (1975). *Design for evolution. Self-organization and Planning in the Life of Human Systems.* New York: George Braziller.

Jantsch, E. (1980). *The Self-organizing Universe: Scientific and Human Implications of the Emerging Paradigm of Evolution..* Oxford, New York: Pergamon Press.

Jung, C.-G. (1953-1979). *Collected Works,* Vols. I-XX, H.Read (ed.), R. Hull (trans.), M.Fordham, G. Adler, and Wm. McGuire. Bollingen Series, NJ: Princeton University Press.

Jung C.-G. (1963). *Memories, Dreams , Reflections.* A. Jaffe, (Ed.). New York: Pantheon Books.

Laszlo, E. (1991).- *The Age of Bifurcation: Understanding the Changing World.* Vol.1, (The World Futures General Evolution Studies). Amsterdam: Gordon and Breach Science Publishers.

Laszlo, E. (1995). *The Interconnected Universe: Conceptual Foundations of Transdisciplinary Unified Theory.* Singapore: World Scientific.

Luhmann, N. (1995). *Social Systems.,* tr. J. Bednarz, Jr. with D. Baecker. Stanford, CA: Stanford University Press.

Merrell, F. (1995). *Peirce's Semiotics Now, a Primer.* Toronto: Canadian Scholars' Press.

Merrell, F. (1999). "Living Signs". *Semiotica* 127, 1/4. Berlin, New York: Mouton de Gruyter, 453-479.

Muller, J. (2000). "Hierarchical Models in Semiotics and Psychoanalysis". In *Peirce, Semiotics, and Psychoanalysis.* John Muller and Joseph Brent, (Eds.), 49-67. Maryland: The John Hopkins University Press.

Nichols, S. (1980). *Jung and Tarot: an Archetypal Journey.* York Beach, ME: Samuel Weiser, Inc.

Noddings, N. (1984). *Caring: A Feminine Approach to Ethics and Moral Education.* Berkeley, CA: University of California Press.

Noddings, N. (1993). *Educating for Intelligent Belief or Unbelief.* New York and London: Teachers College, Columbia University.

Noddings, N. (1998). *Philosophy of Education.* (=Dimensions of Philosophy Series). Boulder, Colorado: Westview Press.
Noth, W. (1995). *Handbook of Semiotics.*(Advances in Semiotics). Bloomington:–Indiana University Press.
Pauli, W. (1994). *Writings on Physics and Philosophy.* Charles P.Enz and Karl von Meyenn (Eds.), tr. R. Schlapp. Berlin: Springer-Verlag.
Peat, David F. (1992). The Science of Harmony and Gentle Action. In *The Interrelationship between Mind and Matter.* Rubik, B. (Ed.). The Center for Frontier Sciences at Temple University, Philadelphia, Pennsylvania, 191-205.
Peirce, C. S. (1860-1911). *Collected Papers by Charles Sanders Peirce*, Charles Hartshorne and Paul Weiss (Eds.). Cambridge, MA: Harvard University Press, 1931-1935.
Peirce, C. S. (1955). *Philosophical Writings of Peirce,.*Justus Buchler (Ed.). New York: Dover Publications.
Penrose, R., Shimony, A., Cartwright, N., Hawking, S. (1997). *The Large, the Small, and the Human Mind..* UK: Cambridge University Press.
Samuels. A. (1985). *Jung and the Post-Jungians.* London and New York: Routledge.
Samuels. A., B. Shorter, and F. Plaut. (1986). *A Critical Dictionary of Jungian Analysis.* London and New York: Routledge.
Sebeok, Thomas A. (1991). "Communication". In *A Sign is Just a Sign.* (Advances in Semiotics), Bloomington: Indiana University Press, 23-35.
Sebeok, Thomas A., Ed. (1994). *Encyclopedic Dictionary of Semiotics.* (Approaches to Semiotics; 73). Berlin, New York: Mouton de Gruyter.
Semetsky, I. (1994). *Introduction of Tarot Readings into Clinical Psychotherapy – Naturalistic Inquiry.* Unpublished Thesis, Pacific Oaks College, Pasadena, CA.
Semetsky, I. (1998). "On the Nature of Tarot". *Frontier Perspectives 7(1),* The Center for Frontier Sciences, Temple University, PA, 58-66.
Semetsky, I (1999a). "Tarot Semiotics as Cartography of Events".-In *Semiotics 1998,* eds. C.W. Spinks and John Deely. New York: Peter Lang Publishing, Inc., 38-51.
Semetsky, I. (1999b). "Self-Organization in Tarot Semiotics". Paper presented at the 7th International Congress of the International Association for Semiotic Studies, *Sign Processes in Complex Systems,* Dresden, University of Technology, October 6-11, 1999. Forthcoming in the Proceedings.
Semetsky, I. (2000a). "The Adventures of a postmodern Fool, or the Semiotics of Learning". In *Semiotics 1999,* eds. S. Simpkins, C.W. Spinks and John Deely. New York: Peter Lang Publishing, Inc., 477-495.
Semetsky, I. (2000b). "The End of a Semiotic Fallacy". *Semiotica* 130,-3/4. Berlin, New York: Mouton de Gruyter, 283-300.
Semetsky, I. (2000c). "Symbolism of the Tower as Abjection". *Parallax 15,*vol.6, no.2, Leeds, UK: Taylors & Francis, 110-122.
Semetsky, I. (2000d). "The Magician – Marketplace Teacher, or Eros contained and uncontained". Paper presented at the 25th Annual Meeting of the Semiotic Society of America, Lafayette, Purdue University.Septmeber 28-October 1, 2000. Forthcoming in *Semiotics 2000.*
Sheriff, John K. (1994). Charles Peirce's Guess at the Riddle. Bloomington: Indiana University Press.
Shimony, A. (1993). *Search for a Naturalistic World View,* vols. I-II, Cambridge University Press.
Spinks, C. W. (1991). "Diagrammatic thinking and the portraiture of thought". In *On Semiotic Modeling.* Myrdene Anderson and Floyd Merrell (Eds.). Berlin, New York: Mouton de Gruyter, 441-481.
Stapp, H..(1993). *Mind, Matter and Quantum mechanics.* Berlin: Springer-Verlag.
Van Eenwyk, John R.(1997). *Archetypes & Strange Attractors: The Chaotic World of Symbols.*(Studies in Jungian psychology by Jungian analysts; 75). Toronto, Canada: Inner City Books.
Von Eckardt, B. (1996). *What is Cognitive Science?.* Cambridge, MA: The MIT Press

ASC
American Society for Cybernetics
a society for the art and
science of human understanding

On the Cybernetics of Fixed Points

Louis H. Kauffman

In his paper "Objects as Tokens for Eigenbehaviours" [2] Heinz von Foerster suggests that we think seriously about the mathematical structure behind the constructivist doctrine that *perceived worlds are worlds created by the observer.* At first glance such a statement appears to be nothing more than solipsism. At second glance, the statement appears to be a tautology, for who else can create the rich subjectivity of the immediate impression of the senses? At third glance, something more is needed. A beginning in that direction occurs with Heinz' paper. In that paper he suggests that the familiar objects of our experience are the fixed points of operators. These operators *are* the structure of our perception. To the extent that the operators are shared, there is no solipsism in this point of view. It is the beginning of a mathematics of second order cybernetics.

Where are these operators and where are their fixed points? Lets start back closer to the beginning. Wittgenstein says, at the beginning of the Tractatus [3],

"The world is everything that is the case." What is the case are the distinctions, including the distinction that there is a world at all. It is tempting to succumb to the idea that behind this tapestry of distinction there is a hidden inner mechanism of the "thing in itself" hiding behind a world of appearances. That "thing in itself" is the other side of the distinction that is world of appearances. One can take the point of view that the world is the world of appearances. But one can take the agnostic point of view that a distinction can be deeply investigated from one of its sides without a belief in the existence of an unobservable side. It is, I believe, this agnostic point of view that leads directly to objects as tokens for eigenbehaviours.

For consider the relationship between an observer O and an "object" A. The key point about the observer and the object is that "the object remains in constant form with respect to the observer". This constancy of form does not preclude motion or change of shape. Form is more malleable than the geometry of Euclid. In fact, ultimately the form of an "object" is the form of the distinction that "it" makes in the space of our perception. In any attempt to speak absolutely about the nature of form we take the form of distinction for the form. (parphrasing Spencer-Brown [1]). It is the form of distinction that remains constant and produces an apparent object for the observer. How can you write an equation for this? The simplest route is to write

$$O(A) = A.$$

The object **A** is a fixed point for the observer **O**. The object is an eigenform. We must emphasize that this is the most schematic possible description of the condition of the observer in relation to an object **A**. We only record that the observer as an actor (operator) manages through his acting to leave the (form of) the object unchanged. This can be a recognition of the symmetry of the object but it also can be a description of how the observer, searching for an object, makes that object up (like a good fairy tale) from the very ingredients that are the observer herself. This is the situation that Heinz von Foerster has been most interested in studying. As he puts it, if you give a person an undecidable problem, then the answer that he gives you is a description of himself. And so, by working on hard and undecidable problems we go deeply into the discovery of who we really are. All this is symbolized in the little equation **O(A) = A**.

And what about this matter of the object as a token for eigenbehaviour? This is the crucial step. We forget about the object and focus on the observer. We atttempt to "solve" the equation **O(A) = A** with **A** as the unknown. Not only do we admit that the "inner" structure of the object is unknown, we adhere to whatever knowledge we have of the observer and attempt to find what such an observer could observe based upon that structure.

The rest of the paper is a multi-logue about the attempts to solve the equation of the obsever in relation to his/her observation. We first encounter Mr. D, who has solved his own equation in such a way that he has no head and instead has a great open space of possibility where his head was supposed to be. This requires a drink to ingest and we go to Zermelo's Bar, where we find two mathematicians arguing over the solution to an equation whose solution is the Golden Ratio, a proportion well known to the Greeks. The mathematicians are a little hard to follow, but their discussion turns on all the essential issues of recursion, reality and infinity that we will need for this adventure. Then Dr. Von F appears in the bar (we think you can guess who this is) and explains the nature of eigenforms. He is followed by a character named Charlie and Dr. CC, a linguist and logician, then by Dr. HM, a biologist. Later there appears a physicist, Dr. JB and finally Dr. R himself, the source of the self-referential paradox. We hope that you will join in on this discussion yourself.

Infinite Recursion and Its Relatives

Our problem is to solve the equation $O(A) = A$ for A in terms of O.

For example, suppose that the observer O is Mr. D, a man who insists that he has no head. We interview him. Well Mr. D, why do you say that you have no head? Mr. D. replies. Oh it is so simple, you will see at once what I mean. In fact, consider what you yourself see. Look directly around. Do you see your head? No. You see and feel a great open space of perception where your head is supposed to be, and a flow of thoughts and feelings. But no head! The body comes in.

Shoulders, arms, legs, shoes and the world. But no head. Instead of a head there is a great teeming void of perception. Once I realized this, I knew that the relationship of a self to reality was indeed deep and mysterious.

As we can see, Mr. D has discovered that what is constant for his visual observer is a body without a head. He has solved the problem of finding himself as a solution of the equation of himself in terms of himself. Perhaps we need a drink.

We walk into Zermelo's Bar and two mathematicians appear on the scene. One says to other: How do you solve this equation? I want a postive real solution.

$$1 + 1/A = A.$$

The second one says: Nothing to it, we multiply both sides by the unknown **A** and rewrite as

$$A + 1 = A^2.$$

Then, solving the quadratic equation, we find that

$$A = (1 + \sqrt{5})/2.$$

The first mathematician says: Nice tricks you have there, but I prefer infinite reentry of the equation into itself. Look here: If **A** = 1 + 1/**A**, then

$$\begin{aligned}A = \\ 1 + 1/A = \\ 1+1/(1+1/A) = \\ 1+1/(1+1/(1+1/A)) = \\ 1+1/(1+1/(1+1/(1+1/A)))\end{aligned}$$

and I will take this reentry process to infinity and obtain the form

$$A = 1+ 1/(1+1/(1+1/(1+1/(1+ 1/(1+1/(1+...)))))).$$

The second mathematician then says: Well I like your method. We can combine our answers and write a beautiful formula!

$$(1+\sqrt{5})/2 = \\ 1+ 1/(1+1/(1+1/(1+1/(1+ 1/(1+1/(1+...))))))$$

Why do you like this formula? says the second guy. Well, sez the first guy, the left hand side is a definite irrational number and it is easy to see by squaring it that it satisfies the equation $A^2 = A + 1$ as we wanted it. But irrational numbers have a curiously tenuous existence unless you know a way to calculate approximations for them. On the other hand, your right hand side can be regarded as the limit of the fractions

$1 = 1/1$
$1+1/1 = 2/1 = 2$
$1+1/(1+1/1) = 3/2$
$1+1/(1+1/(1+1/1)) = 5/3$
$1+1/(1+1/(1+1/(1+1/1))) = 8/5$
$1+1/(1+1/(1+1/(1+1/(1+1/1)))) = 13/8$
$1+1/(1+1/(1+1/(1+1/(1+1/(1+1/1))))) = 21/13$

with the first few terms of this limit being

$$(1+\sqrt{5})/2 = 1.618...$$

On top of this your infinite formula actually does reenter itself as an infinite expression it really is of the form
$$A = 1 + 1/A.$$
The first guy comes back with: Well it sounds to me like you really believe in the "actual" infinity of the terms on the right-hand side. I also like to imagine that they are all there existing together in space with no time.

Right! says the second guy. We know that this is an idealization, but it lets us reason to correct answers and to put them in an aesthetically pleasing form.

The bartender is listening to all of this, and he leans over and says: You guys have to meet a couple of others on this score. There is Dr. Von F and Dr. CC. They both have some ideas very similar to yours. Hey, here is Dr. Von F now. Dr. Von F, could you tell these fellows about your eigenforms?

Jah! Of course! It is all very simple. We just combine this notion of recursion with the most general possible situation. Suppose we have any observer O and we wish to find a fixed point for her. Well then we just let the observer act without limit as in
$$A = O(O(O(O(O(O(O(O(...))))))))).$$
After infinity, one more application of O does not change the result and we have
$$O(A) = A.$$
This is very simple, no? And it shows how we make objects. These objects are the tokens of our repeated behaviours in shaping a form from nothing but our own operations. As I have said before, the human identity is precisely the fixed point of such a recursion. "I am the link between myself and observing myself." [2]

The first mathematician makes a comment: What you are doing is a precise generalization of my infinite continued fraction! If I had defined
$$O(A) = 1 + 1/A$$
then we would have
$$O(O(O(...))) = 1+ 1/(1+1/(1+1/(1+...))).$$
But I am puzzled by your approach, for it would seem that you are willing that your solution A will have no relation with how the process starts, and also it may not be related to the original domain in which it was constructed!

For example, in my mathematics, I could consider the operator
$$O(A) = -1/A$$
and this operator does not have a fixed point in the real numbers, but if we take $A=i$ where $i^2 = -1$ (the simplest imaginary number), then $O(i)=i$. Are you suggesting that
$$i = -1/-1/-1/... ?$$
Dr. Von F replies: Jah, Jah! This is very important! The fixed point can be a construction that breaks ground into an entirely new domain! Actually, I am mainly interested in those fixed points that do break new ground. We are looking for the places where new structures emerge. In your mathematics you have illustrated this in two ways. In the first recursion, the values converge to an irrational number (the golden ratio). All the finite approximations are rational

fractions (ratios of Fibonacci numbers) but in the limit of the infinite eigenform, you arrive at this beautiful new irrational number! And in your second example all the finite approximations oscillate like a buzzer, or a paradox, between positive unity and negative unity, but the eigenform is a true representative of the imaginary square root on minus one! And don't forget that this "imaginary" quantity is fundamental to both logic and physics. The fully general eigenforms are fundamental to the ontology of the world.

Suddenly the door to Zermelo's Bar opens and in walks a character that everyone calls "Charlie." Charlie! says the barkeep, where have you been? We have a good discussion on signs going here. You have to hear this stuff. Charlie says, Well I heard just about everything Dr. Von F said as I admit here to a bit of eavesdropping on the other side of the door! These eigenforms of Von F are quite familiar to me as I have thought continuously along these lines for many years. You see, any sign once you look at it in the context of its reference and the continous expansion of its interpretant becomes a growing complex of signs referring to other signs, growing until the references close on themselves and, as Dr. Von F correctly describes, these closures are the eigenforms, the tokens for apparently stable behaviours. As the complex of signs grows, the complex itself is a sign and as the closures occur that sign becomes a sign for itself. We humans are in our very nature such signs for ourselves.

Dr. Von F says: Well I always say that I am the link between myself and observing myself. I am a sign for myself!

At this point Dr. CC chimes in: But Dr. Von F and Charlie, this excursion to recursion and infinity seems quite excessive! It is all right for mathematicians to imagine such a thing, but we humans exist in language and the finiteness of expressions. Surely you do not suggest that this profligate composition of the operator and expansion of sign complexes actually happens!

Well, Dr. CC, says Von F, I am really a physicist and well aware of the speed of physical process in relation to the very slow pace of our verbal thought. Surely you have stood between two facing mirrors and seen the near-instantaneous tunnel of reflections created by light bouncing back and forth between the mirrors. Yes, I am seriously suggesting that the self-composition of the observer is carried to high orders. These orders are sufficiently large and accomplished with such a high speed that they appear infinite in the eyes of the observer. Now you may detect the beginning of a paradoxical flight here. The very observer who is too slow to detect the difference between a large number and infinity is yet so quick and subtle that he/she can produce this flight to infinity. But I beg your pardon, this is still a matter of the interaction of slow thought and fast action. Wave your arm back and forth rapidly in front of your eyes. For all practical purposes the arm appears to be in two places at the same time! You do not deny that it is "you" that moves the arm, and it is "you" that perceives it.

I simply go further and suggest that every perception is based on such an illusion of permanancy, based on the self composition of your self. You do it all and you are surprised at the result — you can not perceive all that you do!

Charlie adds: I agree but do not have to rest on physics. Our shortsighted view of our own nature arises from the difficulty in reckoning that our true nature is as signs for ourselves. It is only at the limit of eigenbehaviours that such signs appear simple. We partake of the complexity of the universe.

Dr CC replies: Ah Charlie and Dr. Von F, I have been working in the linguistic and logical realm and you will see that our points of view are mutually supporting. For I imagine the structure of the observer as a big network of communicating entities. These entities have so much interrelation among themselves that their identities begin to merge into one identity and that is the apparent identity of the self.

Charlie interrupts with: Yes! That is the essence of continuity.

Dr. CC continues. I agree! The infinity in my view is not with any one of them, but with the aggregate of them that has become so large as to begin to merge into a continuity.

But let me explain: If A and B are entities in my "community of the self", then they can interact with each other and with themselves. These processes of interaction produce new entites who exist at the same level as the original entities. Can you imagine this? Of course you can, you are such an entity. For example, I suggest to you that you are the self that thinks kindly of others, that you satisfy the equation **SX = KX** where **S** is "you" and **KX** is the being "thinks kindly of **X**". *Then that entity **S** exists.* In the world of language, *every definable entity exists.* The consequence is that **S** might even think kindly of herself as in **SS = KS**. That **S** can think kindly of herself is, in this linguistic world, dependent on the condition that the kindly thinking observer is an observer at the same level as any other observer. Now there are many such entities. Watch this magic trick. Let

$$GX = O(XX).$$

The entitity **G** is the observer who observes an entity observing herself. What happens when G observes herself? Then **G** observes herself observing herself and we have a fixed point, an eigenform!

$$GG = O(GG).$$

I have constructed the eigenform without the infinite composition of the observer upon herself. Of course once this self-reflexive construction comes into the being of language then it runs automatically to the level of practical infinity and produces your recursion.

$$GG = O(GG) = O(O(GG)) = O(O(O(GG))) = \ldots$$

I believe my linguistic construction provides the context for your observer's self interaction. The true infinity in my world is a distributed infinity of beings each coming into being as a name for a process of observation. This contines without end and is the basis of the coincidence of the language and the metalanguge in this world.

At this point Dr. HM, a biologist, walks into the room. He remarks: I see that you have been discussing the stability of perceptions from physical and linguistic principles. Let me tell you how I see these matters in my domain. The beings you talk about are biological, not just logical. They exist in the evolutionary flow of coordinations of coordinations that give rise to the mutual patternings that you call "language" and "thought". It is not at all surprising that each such being, coordinated with the others in the deep flow of its history in biological time will appear layered like an onion with the actions of each on each. The long time history of mutual interaction and coordination will generate the appearance of the eigenforms. But there is no "disembodied observer" who generates these forms from some abstract place. In biology there is no problem of mind (abstract observer) and body. They are one. Mind and observer both refer to the conversational domain that arises in the construction of the coordination of coordinations that is language. The disembodied observer is a fantasy that is convenient for the mathematician or the physicist. In the biological realm all forms are generated through time in an organic way.

And finally, Dr. JB enters the room, a very theoretical physicist. He says: Ah it is not surprising, but you all have the business of objects and eigenforms quite wrong. Let me start with the views of the biologist Dr. HM. You see, there is no time. None. Time is an illusion. Of course in order to tell you about this insight I shall have to use words that appear to describe states in time. That is my fate to be so projected into language. You must forgive me.

Each moment of being is eternal, beyond time. I prefer to call such moments "time capusules." Each moment contains that possibility that it can be interpreted in terms of a "history", a story of events leading up to the "present moment" that constitutes the time capsule as a whole. But this history is a pattern in eternity. That the history can be told with some coherence and that we manage to tell the story of "past events" leads us to believe that these past events "actually happened". But in fact what has happened is happening now and only now in the eternity of the time capsule whose richness dervives from the superposition of its quantum states.

At this point the bartender chimes in: I'll drink to that. Time is a grand illusion and a wee scotch from my bar will convince ye o' that in less time than it takes to wink an eye!

All well and good, says Dr. R, who just walked into the bar, but as I was telling my friend Frege, if there is one thing that will give us trouble it is this notion of eternity and the non-existence of time. For as I told Gottlob just the other day, you have only to imagine the timeless reality of the set of all sets that are not members of themselves and you will have to leave logic behind! I gave up long ago my travails on this issue with Professor Whitehead. We tried to make logic go first and it was a disaster. Now I let logic run along behind and there is no problem at all. As far as fixed points are concerned my favorite is Omega, the set whose only member is Omega herself. You see that the act of set formation is nothing but an act of reflection. Omega finds herself in reflecting on herself.

Dr. CC retorts: Well, Russell, I hardly expected you to capitulate your position on logic. Your Type is hardly likely to just slip away. I prefer to make a specimen of your famous set in the following way. I let **AB** mean that "B is a member of A".

Then I define your set of all sets that are not members of themselves" by the equation

$$Rx = {\sim}xx.$$

Then we can pin the specimen to the board by substituting **R** for **x** as in

$$RR = {\sim}RR.$$

This **RR** is a fixed point for negation. It is neither true nor false. I do not leave logic behind. I imagine new states of logical discourse that are beyond the true and the false. Your set performs this transition to imaginary Boolean values.

Now Dr. HM says: Well I see you fellows are beginning to foment an argument. I feel that I must point out to you that logical paradox occurs only in the domain of language. There is no such matter as the paradox of the Russell set in the natural domain. In the natural domain, all apparent contradictions are only antimonies in the eyes of some observer. Nature herself runs in the single valued logic of the evolutionary flow. This is why I emphasize that it is only in the linguistic domain of coordinations of coordinations that the eigenforms arise. At the biological level there are processes that can be seen as recursions, but this seeing is already at the level of the coordinations. There is no mystery in this, but it is neccessary to round out the mathematical models with the prolific play and dynamics of the underlying biology.

In this sense biology is prior to physics as well as cognition.

At this point a tremour shakes the bar and the lights go out. I am sorry folks, the bartender says from the darkness, but this is another one of our natural events in the single valued logical flow of biological time — a small earthquake. I will have to ask you to leave now for your own safety. And so the discussion ended, unfinished but perhaps that was for the best.

Notes and a Mathematical Appendix

The story in section 2 presents a number of different points of view about the cybernetics of fixed points. Fixed points can be produced by infinite recursion, by direct self-reference, through the linguistics of lambda calculus, and by approximation to infinites. Mr. D is a fictionalized version of Douglas Harding the man who indeed realized that he did not have a head, and had the courage to write about it. The good Drs. at the bar represent these points of view and are thinly disguised representatives of the viewpoints of Heinz von Foerster, Alonzo Church and Haskell Curry (Dr. CC), Humberto Maturana and the physicists Julian Barbour. Charlie represents the American mathematical philosopher Charles Sanders Peirce. Dr. R. represents Bertrand Russell, the inventor of the set of all sets that are not members of themselves. All this is only the beginning. The most famous fixed point of them all is the Universe herself, acted here by the bartender.

References

1. G. Spencer-Brown, "Laws of Form", George Allen and Unwin Ltd. London (1969).
2. H. von Foerster, Objects as tokens for eigenbehaviours, in "Observing Systems", Systems Inquiry Series (1981), pp. 274-285.
3. L. Wittgenstein, "Tractatus Logicus Philosophicus", Routledge and Kegan Paul, London & NY (1922).

A (Cybernetic) Musing:
Constructing My Cybernetic World.

Ranulph Glanville[1]

Introduction

I am going to use the occasion of the arrival of the (proper-ish) new millennium to write about aspects of my own work. I hope this won't seem very self-centred or indulgent. The idea came to me at Christmas: like many, I send relatives and friends a 'Year Report'. This year I decided, also, to attempt to explain what I've been working at in cybernetics for the last 30 years, or maybe my whole life.[2] I know that what I wrote (and present here in reworked form) is under-argued and it might be relatively easy to pick holes in it. I hope you'll not want to do that: this is an attempt to set a personal second order cybernetic scene which was originally intended for the completely uninformed, rather than a technical piece. On friend commented (about the Christmas version) that not only was it about a constructed world, but it actually constructed that world as well.

What I believe ties together this work is the belief that each of us experiences and understands the world we find ourselves in in our own way: that is to say, each of us is both distinct and different (I am not arguing here about the nature of these worlds). And the attempt is to find the conditions under which we might begin to explain how we could understand the world. It is not philosophy, science or psychology: rather I think what I've been doing is attempting to lay the ground so the philosophers, scientists and psychologists can do their work. I am thinking about thinking: that is, reflecting.

I am explaining experience, and the experience you have is not mine, nor even of that which I believe I experience; but is of my account and my explanation. My explanation is where I am now. I cannot regress to that state where I was what we as informed adults call new-born, when I was, perhaps, a tabula rasa. I cannot go behind the concepts I've formed, or the fact that I form them, or the fact that knowing this, I am where I am now, who I am now. The explanations that I make change me. What they are based in is my past, so they are cumulative. No wonder that psychotherapy requires such constant practise, unless one is blessed by that

---

[1] Independent Academic, CybernEthics Research, 52 Lawrence Road, Southsea, Hants, PO5 1NY, UK. tel +44 23 92 73 77 79; fax +44 23 92 79 66 17; email ranulph@glanville.co.uk.

[2] What I wrote was also occasioned by the publication of the posthumous autobiography of my uncle, Frank Tindall (*Memoirs and Confessions of a County Planning Officer*, Pantile Press, Ford, Midlothian). I discovered just how much he had done, and how little of it I had realised. I was appalled at my ignorance. So I decided to account for myself.

total reformat-in-one that Saul seems to have undergone on the road to Damascus. I cannot go behind where I am now to where I might have been, without being here, where I am now, to go behind. Even the notion of behind is formed in this way. This is the recursion of life.

To some, I may seem anti-science: I'm not! What I am anti is the misrepresentation of the status of what science can bring us in the way of what we know. Thus, for instance, I dislike the confusion of the description with the thing described, and the (mechanism of the) explanation with the mechanism of that thing.

In a recent piece published on the web, Ernst von Glasersfeld reminds us:

> Constructivism is an epistemological model and can no more be empirically refuted than Leibniz's Monadology, Nicholas of Cusa's theory of 'docta ignorantia', or any other theory of knowing. They are conceptual networks built on assumptions which one may or may not like; they have no truth value; what matters is their internal coherence and whether we find them useful.

That he is talking about constructivism rather than second order cybernetics (is there a difference?) is not, of course, the point. What matters is that it is understood that I am not trying to do science, or anything like that: I am trying to set up a system within which we can have such as science.

To start with

I believe in the individual, and in the distinction between each of us: I am convinced that each of us is different, and this difference is important. We cannot demonstrate we are or think the same: even attempting this requires we assume we are different, in order to be able to find this sameness. I have noticed that we tend, in finding similarities, to forget the differences: the majority of discussions of our experience are focused towards a knowledge based on what we hold in common, ignoring difference. In contrast, I am less interested in what we have in common than what we have that supports and keeps us different. I hold this difference to be self-evident. My question is how can we account for a world which each of us sees differently, and which, as a result, we cannot be sure is the same world?

As my response to this I built a framework in which I designed a structure to support difference based in observation (a general term meaning more than visual looking!). The theory is built around what I call 'Objects', which are taken to be 'self-observing'. I know this sounds odd: if it worries you, see the footnote.[1] Objects are taken to have two roles: self-observing and self-observed. But each Object is just one Object, so these roles are seen as switching, which they do by generating time (they are oscillators). When the Object is self-observed, a slot is

---

[1] How do we know something is observable? When it is observed. What does it need, that something may be observed? An observer. What is the minimum configuration for such observing? When the observing is of the self: the observed is the observer. This organisation makes for autonomy and organisational closure.

left open for observing which other Objects may look into (providing they are in their self-observed role, and so are free to observe): an Object can observe another Object by occupying that Object's 'observing' slot while it is empty, which it is when the (self) Object is not in the (self) observing role, but in the observed role. Each Object generates its own time, which means their times appear different to each other, and also that one Object might observe several different, other Objects simultaneously, allowing observations of different Objects to relate several Objects together through (the synchronising times of) our observations of such Objects. Which is the motivation for this work: I wanted to set up a structure allowing that each of us see differently yet believe we see the same. I don't believe anyone else has put it this way: usually differences are to be explained away, not celebrated! (I'm sorry this account is a little dense and garbled.)

What I'm making is an explanation. It is not what is (which I claim we can never know — we can't even know if there is anything separate and independent from our observations (if there is anything to be explained) so if we can know, we can't know that we know!). A serious defect of science as currently, generally recounted occurs because we fail to distinguish between explanation and mechanism. Science is presented in such a manner that explanations are said to be what (actually) happens. This elementary error is made by even the eminent, and by many who should know better, and that is what I am anti![2]

The explanation I constructed, in my Theory of Objects (my first PhD), does all sorts of unexpected things. For instance, explanations of how we develop qualities such as memory and consciousness; the way that arguments don't seem to reveal (lasting) fundamentals; and how we represent and communicate — without which there would be no point in this thesis for, even if we did all see differently (and, therefore, as different), without communication we'd never know it.

Why did I do this? I was accused, at the time, of treating humans as machines. But that's not it — in fact, quite the opposite. I was trying to create an account that leaves us liberated (and I believe I succeeded): the prerequisite of my thesis is that we understand differently. We each see the world in our own way(s). We are free to see as we see. Our mission is to be where we are. Therefore, we are responsible for what we see, and what we do based on this: what we learn and all our actions. This contrasts greatly with other views, where we are slaves to mechanism and without responsibility. It is this view of what humans are, of freedom and the liberal libertarian that has always been my belief.

What I like most about the theory is its terse elegance and power. When I look at it nowadays, I am overwhelmed by it. I like, also, that it does more than satisfy the prerequisites I'd set, and that the structure it makes shows how it is possible to consider a world in which we all see differently, and the consequences of this. In contrast, post-modernism only demands we accept that we do all see differently

---

[2] I am aware that, for some theorists, specially mathematicians, there is a notion that (for instance) the world IS mathematics: mathematics is both explanation and mechanism because the two either map perfectly onto eachother or are one. This is a very cybernetic idea, and one that post-Goedel mathematics would seem to preclude. I think maths and cybernetics are both a bit confused here!

(that any way of looking is as good as any other), and then despairs of finding communication and structure.

Delineating what we might see differently

Since I start from an interest in a structure allowing us all to observe differently while we believe we observe the same thing, I must be interested in how we might describe how we might delineate or identify (so we have an observed). This is important because we do believe we distinguish one thing from another (and that we can talk about this to each other). It is this act which allows us to treat the world as populated with observable Objects: it allows us to believe there are Objects to observe.[3] It is how we either put or observe the lines/boundaries/edges round things. Such lines, so elegantly discussed by the artist Paul Klee (a drawing is 'taking a line for a walk'), have been a lifelong obsession of mine.

There is a key conceptual text which cyberneticians like me refer to: George Spencer Brown's 'Laws of Form', which starts from the primitive act of drawing a distinction. Louis Kauffman's delightful column often touches on this, and there was a wonderfully helpful tutorial in vol 6 no 4. I hope to return to this on a later occasion.[4] Lines don't only distinguish an inside from an outside, they also have qualities of their own.

We demarcate where we think 'things' begin and end by drawing a distinction — by making a line. I'm talking metaphorically: we don't usually pick up a pencil, and 'things' may be very loose, woolly and invisible — they are Objects: making the line may actually create these things (as I, amongst many others, would argue). The line forms an inside (where the 'thing' is) and an outside (where the 'thing' isn't but some other 'thing' is). But we have also created the line, and the line is also a 'thing', so there's a question of how the line is distinguished from what's inside it. To me this suggests that either we must go on distinguishing and re-distinguishing the 'thing' (i.e., we keep on seeing it as the same, or we have an 'edge' of great richness: this we find in, for instance, highly articulated walls in architecture); or the line is what's distinguished (so there's no inside or outside, just the line — pictorially, a Moebius strip rather than a circle). Surprisingly, this understanding seems to match very closely the way that the Mayans thought of their architecture, where they talked of the space of/in the wall. (I discuss this, and the work I did on how we perceive architectural space, in my second PhD.)

---

[3] Elsewhere, I've argued how we can take observations and turn them into credible 'real-world objects': I'll not do that here (though I do precis the argument later in this article), even though it is very important, for it explains how, within the way of thinking I'm putting forward, we can have a common, negotiated 'reality of experience', and hence the science and technology we all so depend on and enjoy. I write about it in 'An Observing Science'.

[4] I had a (student) piece published the year before 'Laws of Form', in which I distinguish the Yorkshire town Kirbymoorside through drawing several boundaries (distinctions) as lines drawn on maps, overlapping and containing the little market town.

I could go on. For instance, we talk of things being stable (including bodies, concepts, relationships) often meaning that they continue to be, but there are two types of stability: static (unchanging, how we often seem to ourselves) and dynamic (changing, how we often seem to others). Stability is determined in relation to something else (a goal, which has also to be stable, although I won't pursue that here). When something seems to be dynamically stable we can consider this goal as being outside the stable thing. Yet for the thing to continue to be, there must be a goal inside, and (unless we are also within) we can't do more than surmise there is such a goal. Second order cybernetics is full of circles like this, and of logical conundrums which are best dealt with, I find, by not being insistent on the priority of one outcome or the other; or by not trying to find a solution at all but just being in (actually, inside!) the argument.

There is another aspect to my interest in lines. Say a line I draw does distinguish (for instance) me from what is not me (that is, I distinguish myself, which would make me an Object). What on earth (!) would be the point of doing this if I didn't distinguish myself from something else? What conceivable point is there in having an I if there is no you or it against which to appreciate my I-ness, to enjoy my self? I maintain there is no point. In other words, the purpose of distinguishing myself is not just to assert there is an I, but to distinguish myself from you or it. When I draw a line, I can think of it as making what comes to be each side of it by the act of drawing it, and each one of the three elements I now have (distinguishing line, inside and outside) in a way distinguishes the other two. This is the purpose of (the first) distinction.

There is, I believe, a corollary to this move, to distinguish myself from you. If I am prepared to assume for myself certain qualities, I must allow that these qualities might also be in or of you (or it). And the qualities I see in you must be conceivable for me. Notice I do not say these qualities are present: only that I should leave space in my thinking so they might be — they are potential. I have introduced this concept in previous columns, and call it the 'Principle of Mutual Reciprocity'. It is important because it gives me an ethical obligation beyond simply accepting responsibility, and it means that, in making myself, I already make (and must respect) you and make you different, while still being able to share qualities. Which means that conversation is possible — the basis of decent behaviour.

So nowadays I don't even think there is a 'thing' for us to see differently. I believe we construct a world through a strange sort of interaction: to say there is an I is to say there is a you, and therefore I always construct not the 'thing', but observations that are realisations of interactions, and which, through a process of iteration, I reify into things (which can give us a reality of reference such as is so powerfully used in science). But that's another matter! Whether or not there are things independent of me I can, of course, never know: I should neither assume there are, nor assume there aren't.

I wrote above about inside and outside, and how important they are. We are different: I can never have your understanding, nor you mine (though I can have my understanding of your understanding, which is precisely not your understanding). In a conversation, each of us can be our self, but we also enter into another entity — the conversation we are sharing in creating. Each of us brings our differences — our uniqueness — to this. While outside you, I can be inside the conversation with you. In a conversation, unless one forces his/her view on the other(s), there is bound to be novelty. The conversation will move in its own way, according to what we come to describe, after the event, as its momentum and logic. Conversations lead to novelty and surprise for each participant (though the surprises may be very different). What interaction gives us is novelty — in the sense (at least) of something that was not of me becoming available to me; and often in a much grander sense.

There are different ways of considering the world depending on whether we place ourselves inside or outside (though we are always inside the system of circularity involved in observing). When I am outside the system I can talk of it as having a vast complexity of possibilities ('variety'): the combinatorics of possibilities, if we don't restrict what we can do, is beyond the computable even if the whole cosmological universe is assumed to be a sub-atomic scale computer operating at the fastest of speeds for its whole life-span, as I have reported before in this column. This tells us that the universe we live in is essentially unmanageable and leaves us different options. One is to close down (which is what dictators do): remove the possibilities and the freedoms we have which generate complexity. Another is to accept unmanageability as wonderful: which gives us endless novelty, a richness beyond our wildest imaginings, magic; and demands from us trust, open-mindedness and generosity, the sorts of qualities I assume we would like to claim for ourselves. I love an understanding that promises me novelty and I try to make sure I set up situations so this is encouraged. Even more, I love an understanding that tells me the qualities I most need are the ones I most admire, as opposed to one that tells me that the 'natural' qualities are also the meanest. Don't you?

How could WE know this?

I have slipped between the pronouns I and we in this account, without answering the question of how, if the world is full of I's, there might be a we.

The first part of an answer concerns our ability to observe. Since we can observe Objects (that's the purpose they were set up for), we can observe Objects no matter whether those Objects are what we might now call fellow human beings, quarks or galaxies, love and fear, the totally imaginary or, even possibly, academic papers. It makes no matter what sort of observable we are dealing with when we want to talk of what constitutes observables (the notion of what sort of observable is a later development). So, first, we both observe and have observables, which is, mainly, what I've been writing about. When we can observe Objects, and make observations more than once, we can determine that the observations are the same (are of the same thing). Thus, we can construct an identity for objects of/as if in the real world, as we do (theory sustains experience). We construct an identity between observations which we understand as some thing, an Object.

Being able to put Objects together in this manner gives us the basis for observing relationships between these things. That is to say, if we can find identity (and, by implication, non-identity), we already have a basic relational framework. I will not, here, show how to extend this to a full blown relational logic: it's complicated, and I believe that I have done enough to show the possibility, hoping you will chose to believe me.

The second part concerns the 'what sort of' question. This requires not only that we can observe Objects (or draw distinctions), but that we can find certain qualities in those Objects that we believe they hold in common. This is not so different from the making of identity between observations of an Object, thus making it an object (of the traditional sort), a thing, a concept, a whatever. It is possible to understand this as finding qualities in things, qualities they might share: which assumes the sort of relational logic just mentioned. These qualities allow us to find similarity in quite distinct things, a similarity that supports their difference and separateness.

We create identities between things that are separate and which it is our intention to keep quite distinct, and we often call this representation. A picture (of a thing), no matter how 'picture perfect', is not that thing. Nor is a word the object to which it is attached, or that it indicates. In representation, we keep the difference between the two clear, otherwise it would not work. This is the reason fundamentals can never be described. To say two things are the same is, equally, to say they are different (otherwise they are not two). To name something as fundamental is to say that not only is there something the name and the object hold in common (indivisibility) but that there is something they do not hold in common. Oh dear! Naming a fundamental stops it being fundamental by conceptually splitting it into the part that is the same as the description and the part that is different. However, representation becomes possible, with which comes not only identification, but also the possibility of communicating.

As of now, these connections are made by you or by me: that is, by just one of us. To communicate (or to believe we communicate) we need more than that: I must believe you have made a connection like I have. This is where Gordon Pask's theoretical conversations enter.

Conversation allows us to create agreement, and hence social phenomena, such as everyday language. Language, in this sense, is a collection of agreements: one of the reasons it is so hard for most adults to learn a new language is that they have to accept the rules of a club of users, when they have already learnt to behave, see and describe the world according to the rules of another club. We can (and do) create languages, but there is no reason to presume languages should be precise, contain meanings, or have precedence: language depends on a series of agreements, known to be agreements. We derive language from negotiation. Conversation is not a linguistic device: language is a conversational device.

It is in this process of negotiation that we may communicate, find similarities, and believe there are agreements. Through negotiation we may arrive at social phenomena. Because these language and other phenomena of social agreement derive from negotiations carried out before we were involved, where our only negotiatory power is to accept what's on offer, there must be differences and difficulties.

And design?[5]

Ross Ashby borrowed a fabulous concept from the physicist James Clerk Maxwell — the Black Box. Ashby even suggested everything might be considered a Black Box. I concur, except I'd say *should* instead of *might*. We live in essential ignorance (a good thing, as indicated) and make explanations from/of observations and experiences, which we then test in a circular action: observe, explain, try out, observe, explain etc. In my mind, the Black Box is a philosophical device which sets up and supports a set of positions intimately connected with difference and ignorance.

For some, especially those who want to know what's inside the Black Box, the notion of ignorance is painful. For others, the Black Box is deeply determining (through psychological conditioning). I don't share these views. For me, when I observe a Black Box, I build an explanation of/through my interaction with it. This explanation is (of) what happens between the Black Box and me. But my explanation of what happens is not an actual mechanism: my essential ignorance remains. I don't have to know what happens in the Black Box (what, if anything, makes the behaviour I experience) — in fact I can't, for that's why I invented and placed it. All I can do is make an explanation of the Black Box through my interaction with it, which I believe accounts for its behaviour in this interaction. (Actually, since I only come to posit a Black Box because of an observed change in behaviour, the only evidence of the Black Box is the changing behaviour I observe through interacting: it may not be there at all. Familiar sounding?)

I believe circular behaviour (interaction) with an unknowable is a good description of what designers do (as opposed to what at least some theorists believe they should do!). This approach allows me to understand design, giving it its due place as a serious activity.[6] Designers go through an iterative and circular process, a sort of conversation with themselves using pencil and paper, ending up with something (a design) which is a token of the circular action. The design can be seen as an embodiment of the process and outcome of a conversation. So the Black Box becomes (ex)PLAN(ation).

---

[5] This section responds to the way I have earned my daily bread over the last 30 years: teaching art, design and architecture. I find an enormous similarity between (second order cybernetics) and design: design is a form of action that is, in my view, deeply cybernetic. In this belief I am not alone. Gordon Pask held firmly to it, and, amongst his dozen successful doctoral students at Brunel University, two thirds (by my estimate) had previously studied architecture.

[6] For a long time, designers thought the were inferior scientists.

Designers will recognise this circularity, although not everything in design is circular. There are problems — specified beforehand — to be solved. But the creation of the previously unknown, which is at the heart of design, is reflected in the explanation I give of the Black Box.

This interactive process with the unknowable leads to further interesting understandings. For instance, we can learn about intelligence without having recourse to such nonsenses as (standard) Intelligence Tests. Intelligence is appreciated through interaction with others. It is in the interaction, not in the performance of isolated tasks, that we find intelligence. Equally, we can discover important features of the interface (person to person, person to machine, whatever): this needs space for interaction to occur in — the space between the unknowable and the person who wants to explain it. Current computer interfaces are are mere action/reaction devices, which I believe is one reason computers seem so dumb and so deeply infuriating!

Since we behave in this manner with eachother, we not only use the Black Box model (possibly unwittingly): we are also all designers, for the making of understandings in this way is what designers do. In my (second order cybernetic) view, humans design their understandings and their knowledge into whatever form it takes in our conception. Design is a basic human activity. And design is second order cybernetics in action: it is, if you like, the application of second order cybernetics, making it usefel. Thus, the strong connection I see between cybernetics and design.

References[7]

Ashby, WR (1956) *An Introduction to Cybernetics*, London, Chapman and Hall

Glanville, R (1975) *A Cybernetic Development of Theories of Epistemology and Observation, with reference to Space and Time, as seen in Architecture* (PhD Thesis, unpublished) Brunel University, Uxbridge, also known as 'The Object of Objects, the Point of Points, — or Something about Things'. Note: a web version is planned.

Glanville, R (1988) *Architecture and Space for Thought* (PhD Thesis, unpublished) Brunel University, Uxbridge

Glanville, R (1990) The Self and the Other: the Purpose of Distinction, in Trappl, R *Cybernetics and Systems '90 the Proceedings of the European Meeting on Cybernetics and Systems Research*, Singapore, World Scientific

Glanville, R (1995) Chasing the Blame in Lasker, G (ed) *Research on Progress — Advances in Interdisciplinary Studies on Systems Research and Cybernetics*, vol 11, IIASSRC, Windsor, Ontario

Glanville, R (1996) Communication without Coding: Cybernetics, Meaning and Language (How Language, becoming a System, Betrays itself), Invited paper in *Modern Language Notes*, vol 111 no 3 April (ed Wellbery, D)

Glanville, R (1996) Robin McKinnon Wood and Gordon Pask: a Lifelong Conversation, *Cybernetics and Human Knowing* vol 3 no 4

Glanville, R (1997) A Ship without a Rudder, in Glanville, R and de Zeeuw, G (eds) *Problems of Excavating Cybernetics and Systems*, Southsea, BKS+

Glanville, R (1997) Gordon Pask, ISSS luminaries section, http://www.iss.org.luminaries

---

[7] A CV including a list of my publications with links ot several of them may be found, courtesy of Alex Riegler, at http://www.univie.ac.at/constructivism/people/glanville/cv.html

Glanville, R (1998) A (Cybernetic) Musing: Variety and Creativity, *Cybernetics and Human Knowing* vol 5 no 3

Glanville, R (1999) Living in Lines, in McLeod, R (ed) *Interior Cities*, Melbourne, RMIT Press

Glanville, R (forthcoming) And He Was Magic, in Scott B and Glanville, R eds, memorial issue to Gordon Pask, *Kybernetes*

Glanville, R (in press) An Observing Science, invited paper for special issue of *Fundamentals of Science*

Glanville, R, Hambury, C and Woolston, G (1967) Kirbymoorside, in Donat, J (ed) *World Architecture 4*, Studio Vista, London

Glasersfeld E von (2001) Commentary 10, Response to Nielsen Commentary 8, Karl Jaspers Forum discussion paper TA31 Science, Abundance and Abstraction, from the web site index at http://www.mcgill.ca/douglas/fdg/kjf

Klee, P (1953) *Pedagogical Sketchbook*, London, Faber and Faber

Robertson, R (1999) Some-thing from No-thing: G Spencer-Brown's Laws of Form, *Cybernetics and Human Knowing* vol 6 no 4

Spencer Brown, G (1969) *Laws of Form*, London, George Allen and Unwin

Pask, G (1973) Artificial Intelligence — a Preface and a Theory, in Negroponte N (1979), *Soft Architecture Machines*, Cambridge, MIT Press

Pask, G (1975) *Conversation Theory*, London, Hutchinson

Searching for the Limits of Semiotics: An Extended Manual for an Extended Scientific Field

Reviewed by Nina Ort[1]

In the German- and English-speaking world there are four fundamental manuals or encyclopaedias of semiotics: the *Encyclopedic Dictionary of Semiotics* by Thomas. A. Sebeok (1986), the *Encyclopedia of Semiotics* by Paul Boussiac (1998); *Semiotik* by Roland Posner et al (1997-99) and the *Handbuch der Semiotik* by Winfried Nöth. Especially in Germany, the two reference books by Posner and Nöth today are part of the basic equipment in the academic work dealing with problems associated with semiotics and sign theory.

Nöth's manual is the completely revised and strongly extended edition of the "Handbuch der Semiotik" of 1985, which at that time has been a huge achievement and was welcomed enthusiastically. Now, 15 years later, an updated version of the manual is published, that can easily compete with the other reference works.

Winfried Nöth intends to present a survey of history and extension of what we call the field of semiotics. The second edition of the "Handbuch der Semiotik" is meant to give a representative and pluralistic survey of different evolutions and trends in semiotics at the end of the 20th century, but also to show connections among them.

With this intention the problem of writing the history and a systematic survey of semiotics becomes obvious: the difficulty to determine where the field of semiotics begins and where it ends.

Today it is possible to reconstruct the doctrine of signs by following the advice of the founding committee of the "International Association of Semiotic Studies" in 1969: semiotics is the general notion for what has been subsumed in terms like "Semiologie" or "Semiotics". Even though semiotic questioning reaches far back into history, Nöth assesses the 1980s as the time, when semiotics was generally accepted as a science and varying systematics of semiotics have been developed. Thereby a base is achieved, from which one can look back into history and reconstruct early conceptions of the doctrine of signs as precursors of semiotics – or even as kinds of implicit semiotics – and outline the different scientific fields which were operating with terms of semiotics.

---

[1] **Winfried Nöth**, *Handbuch der Semiotik – the second edition*. Reviewer: Dr. Nina Ort, Ludwig-Maximilians Universität, Institut für neuere deutsche Literatur, Schellingstraße 3, 80799 München, Germany. E-mail: nina.ort@germanistik.uni-muenchen.de

However, signs are observed by such differing disciplines as logics, linguistics, philosophy, biology, psychoanalysis, sociology and a general uniform status of semiotics as a science is not yet validated: Semiotics is assessed to be science, method, fashion, movement, theory, meta-theory or even ideology. Therefore it is a severe problem to structure a manual of semiotics.

Considering the extension of the huge field of semiotics it is necessary to narrow down the choice or to restrict oneself to the demands of a special target group. Nöth therefore decides to create an introductory manual.

In so far, Nöth's manual is organised evident and sensible. It is divided into history and systematic of semiotics, the introduction of classical theorists and the presentation of different applications of semiotics. The new edition has been divided in almost twice as much chapters; the instructive chapter about classical theorists has been supplemented.

Nöth starts with a summary of all historical attempts that have been made to design a theory and a notion of signs. (p. 1-57^2). Therefore he begins – of course in summariness – with Antique and the Middle Ages, mentioning interesting details such as that the antiquely and medieval thinkers set the concept of signs in close context with logic.

In the 17th century the most important precursors of what we call semiotics today were René Descartes, the semiotics of Port Royal, Gottfried Wilhelm Leibniz, Francis Bacon, Thomas Hobbes, John Locke, George Berkeley. In the 18th century E. B. de Condillac, Immanuel Kant and J.G. Herder were playing extraordinary roles by their efforts to draft sign theories. Giambattista Vico is mentioned in a special paragraph. Semiotics in the 19th century was shaped by thinkers as G. W. F. Hegel, W. von Humboldt and B. Bolzano. Finally, Nöth explains the influence of E. Husserl, E. Cassirer, the Marxist Semiotics, Max Bense, T. A. Sebeok, Claude Lévi-Strauss, Jacques Lacan, Michel Foucault, Jacques Derrida and Jean Baudrillard in the 20th century.

The classical theorists of semiotics are introduced in the second chapter about doctrines of semiotics in the 20th century (p. 59-130). Here Nöth presents the works of Charles Sanders Peirce, Ferdinand de Saussure, Louis Hjelmslev, Charles W. Morris, the Russian Formalism, Roman Jakobson, Roland Barthes, A. J. Greimas, Julia Kristeva and Umberto Eco.

I was somewhat astonished at Julia Kristeva being called a classical author of semiotics. As Nöth describes, she develops the notion of "Semanalyse" to designate her "hybrid semiotics", but she has neither elaborated the "Semanalyse" nor founded an own doctrine (see p. 120). I see Kristeva especially in the tradition of the "Psychosemiologie" by Jacques Lacan; therefore I would rather call Lacan a classical than Kristeva.

The third chapter (p. 131-226) is entitled "sign and system", but actually it explains – besides pivotal terms of semiotics – an abundance of notions that occur in sign theory. It is an introduction of different forms of classifying semiotic terms

---

[2] All page numbers refer to the second edtion of the „Handbuch der Semiotik", see references.

such as "sign", "sign vehicle", "semantic", "meaning", "representation", "code" and so on and their different definitions and descriptions. In this chapter, Nöth follows primarily the concept of Ch. S. Peirce, which provides the basics against which other concepts or notional differences stand out. This intensified focusing on the work of Peirce is an improvement on the first edition of the manual.

In the fourth chapter (p. 227-292) Nöth explains the notion of "semiosis". Since semiosis is defined as the process in which a sign develops its effects, it is necessary to explain terms like "communication as a process", "cognition" and "intention". Here again, Nöth primarily refers to the thoughts of Ch. S. Peirce, but, he moreover outlines basic ideas of systems theory about communication (e.g. the triple selection: information, message, understanding), the relevancy of self-reference and the concept of autopoiesis, that has been developed in (radical) constructivism.

Actually, the field of semiosis is outreaching the field of language and linguistic signs, as Nöth shows. He summarises physio-semiotics, eco-semiotics, bio-semiotics, zoo-semiotics; he lines out the evolution of semiotics and drafts the relevancy of time and space regarding semiotics.

After the historical and systematical definition of semiotics, Nöth demonstrates its application in different scientific fields. At first he mentions forms of nonverbal communication (293-322), such as bodily communication, gesticulation, mimic or tactile communication, but also "Proxemics", that means territorial behaviour (semiotic aspects of territory in nonverbal communication), and "Chronemics", the relevancy of time in the nonverbal communication and the social and cultural life.

The best known fields of semiotic science are language and codes of speech (p. 323-390). Linguistic signs, language as a sign system and communication as semiosis are still the crucial elements of semiotics. Nevertheless, linguistics is assessed to be the actual science of language. Nöth therefore describes the relationship between linguistics and semiotics, pointing out, that it is controversial, whether semiotics is superior to linguistics or vice versa, or both are complemental.

Nöth pays much attention to explain different ways of conceptualising arbitrariness and convention, metaphor, text, but also universal, para-, sign language and language substitutes.

After that he focuses on text-semiotics (p. 391-424) whereat text-semiotics is understood as including for instance literature semiotics but not excluding "parole", spoken speech. Some authors therefore prefer the term "discourse". Text semiotics partially coincides with traditional topics such as rhetorics or stylistics; rhetorics for instance appear in the pragmatic dimension of semiotics.

Narratology is semiotic text theory that deals with text forms such as narratives, myths or even ideology. Nöth describes text semiotics as contiguous to hermeneutics but it is questionable if both can be compared with each other.

The manual dedicates an own chapter to aesthetics as literature semiotics (p. 425-466). Here, Nöth respects semiotics of music as well as semiotics of pictorial arts, architecture and theatre.

In this chapter he also re-conceptualises the crucial ideas of Russian Formalism, "poeticity" and "literarity" of literary texts.

The two final chapters deal with media semiotics (p. 467-512), semiotics of culture (p.513-538), social semiotics and interdisciplinary extensions. Here the reader gets information about semiotic aspects for instance in comics, film, magic and even daily life. Media- and cultural semiotics are new aspects, which have been added to the second edition of Nöths's handbook.

The volume is concluding with an encyclopaedic bibliography of nearly a hundred pages, an index of proper names and a subject index.

The advantage of this way of organising the huge field of semiotics is the possibility to discern connections between very heterogeneous problems of semiotic subjects in the different semiotic approaches and applications. The volume invites to investigate those interconnections.

In addition, Nöth does not limit his descriptions to the particularization of bibliographical references that interpret the semiotic aspects of the actual authors, but also gives hints on general introductory works. Since most of the researchers of semiotics are working in very special contexts, general introductory books to their particular works are very helpful. For example, Nöth mentions the French psychoanalyst Jacques Lacan (p. 49-51), who has developed a concept of the psychical apparatus as a sign-based and semiotically structured complex (for this reason Michael Wetzel has chosen the term "Psychosemiologie" to designate the work of Lacan), and that is resembling the concept of Ch. S. Peirce amazingly because of its genuine triadic structure. In his case, for instance, it is of advantage, that Nöth also indicates some introductory works to the psychoanalysis of Jacques Lacan.

The increasing relevancy of (sign) theories that deal with problems of auto-reflexivity and self-reference, such as systems theory, second order cybernetics and constructivism, is discussed in chapter III, *Sign and System* in only six pages. Here, Nöth does not confine his description to these models but also mentions the systems theory's idea of language and literature as systems and cybernetic principles – first order cybernetics in this context (that means self controlling processes of communication). This paragraph seems somehow short, since systems theory, second order cybernetics and constructivism pursue Peirce' idea of genuine triads in a way that lets assume serious consequences in general theory designing. In the scope of these theoretical models the question for a definition of the semiotic sign is still discussed. Auto-reflexivity as constituent of semiotic signs does not only lead to questions of meta-levels (as Nöth exemplifies for instance with Luhmann's model of "Erwartungserwartungen" – expectation of expectations – in communication; see also Watzlawick :"You can not *not* communicate!"; see p. 240).

Moreover it leads theory design to fundamental paradoxes due to the underlying two-valued logics (see for instance: Niklas Luhmann: *Sign as Form*. In: Dirk Baecker, ed., Problems of Form. Stanford: Stanford UP, 1999, pp. 46-63). Paradoxes as a result of trying to found theory architectures on models of differentiation are therefore the crucial problem of theory design and play an extraordinary role in sign theories, too.

Auto-reflexivity raises the question, whether two-valued sign-models are sufficient to describe semiosis, i.e. processing signs. Put it the other way round: the assumption of a triadic sign suggests to mould signs as processing signs – as genuine semiosis.

The meaning of the discussion about the bivalence or trivalence of semiotic signs seems to be somehow underemphasized in Nöth's manual (he just introduces trivalent sign models in pages 136-141). For auto-reflexivity must be designed as a triad. Auto-reflexivity can be assessed to be an evolutionary step of theories; or, as Oliver Jahraus puts it: only when having become auto-reflexive, theory comes to itself.[3]

Under conditions of the universality and auto-reflexivity of signs, semiotics has the chance to become a general theory (such as systems theory, second order cybernetics and constructivism). Generality and auto-reflexivity render these theories compatible, and in research they *defacto* are conflated.

Insofar, semiotics can be designated as trans-disciplinary tool that can be applied in all kinds of scientific fields. It is exactly this point of view that allows second order cybernetics, constructivism and system theory not only to describe semiotic problems by referring to e.g. logics and philosophy, but also to develop semiotics in this trans-disciplinary way.

Observing and processing semiotics as a trans-disciplinary tool then requires, in my opinion, to mention those thinkers, who today have big influence on the design of the notion of signs, such as George Spencer Brown (*Laws of Form*), Gotthard Günther (*Non-Aristotelian Logics*) and Ranulph Glanville (*cybernetics*).

In that respect I would object to Nöths manual. Besides the historical parts, it is dealing mainly with applications of semiotics in various scientific fields, but for the most part neglects the trans-disciplinary constitution of semiotics as a tool. Semiotics as a self-referential science must be applicable on itself, it must be both – science and meta-science (as Nöth mentions right in the preface to the book; see p. XII). Therefore I assess the trans-disciplinary status of semiotics and the trans-disciplinary efforts to establish semiotics as a theory to be as important as the applications of semiotics in different scientific fields.

On the other hand Nöth mentions trans-disciplinary efforts in a special case: his manual is the first one, that integrates the concept and notion of cybersemiotics by Søren Brier as synthesis of systems theory, second order cybernetics and the semiotics of Ch. S. Peirce (p 215).

The aspect of trans-disciplinarity has been put prominent in the manual by Roland Posner. There, semiotics is observed as general and uniform phenomenon, that connects all forms of living nature, social and cultural life. In his manual semiotics is defined as "inter-disciplinary" science.

---

[3] Jahraus, Oliver: Wie verhalten sich Luhmannsche Systemtheorie und Peircesche Zeichentheorie zueinander? In: Jahraus/Ort (in print) (ed): Kommunikation – Bewußtsein – Zeichen. (=STSL), Tübingen, pp. 245-252, here p. 249

Both manuals can not easily be compared. In three volumes Posner assembles the contributions of 175 authors – all of them specialised in certain fields of semiotics. Consequently, the manual has a completely different systematic and a different objective target. In a first part Posner's manual discusses the system of semiotics (emphasizing the notion of "semiosis", which I consider useful), the large second part is working out all sorts of semiotic problems by following a historical timeline. Posner's handbook consequently adresses experts in semiotics in the first place. Only those readers, versed in semiotics, can foresee for instance, what they will get, when reading a contribution by Umberto Eco.

In a way, Posner's manual is a collection of interpretations of semiotics; compared with this, Nöth presents a coherent introductory book to the fields of semiotics and rather information than interpretation.

The second edition of Nöth's manual is an important assistance for students at university who need extensive and expository treatises. It will serve as a steady reference volume for those, who want to extend their knowledge or want to explore, how semiotics is applied in neighbouring disciplines. Right in this option one can see the chance, that semiotics will continue to develop to a trans-disciplinary theory.

References

Boussiac, Paul (1998) (ed.) Encyclopedia of Semiotics. New York: Oxford University Press.
Nöth, Winfried (2000) Handbuch der Semiotik. 2. revised and extended edition. J.B. Metzler: Stuttgart, Weimar. 667 p. DM 78,—; EUR 39,88.
Posner, Roland et al (1997-99) Semiotik: Ein Handbuch zu den zeichentheoretischen Grundlagen von Natur und Kultur. 3 vols. Berlin: de Gruyter.
Sebeok, Thomas A. (1986) Encyclopedic Dictionary of Semiotics. 3 vols. Berlin: Mouton de Gruyter.

Biosemiotica[1]

Winfried Nöth[2]

With this special issue on biosemiotics, the general editor of *Semiotica*, Thomas A. Sebeok, pursues once more his editorial guideline "to encourage the growth of emerging sub-domains of semiotics" (p.1). The volume consists of two parts, *Biosemiotica I* (pp. 5-131), edited by Sebeok himself, and *Biosemiotica II* (pp. 133-655), with the Copenhagen biologists Jesper Hoffmeyer and Claus Emmeche as guest editors. The first part contains five papers on biological aspects of sign processing according to some classics of modern semiotics; the second assembles 25 papers on topics at various interfaces between biology and semiotics.

The classics studied in *Biosemiotica I* are Charles S. Peirce, Victoria Welby, Charles Morris, Roman Jakobson, and Yuri Lotman. Sebeok, who might himself have been included in this genealogy, suggests that Susanne K. Langer and Ernst Cassirer are 20th century philosophers who also belong to the prehistory of biosemiotics. The special relevance of Jakob von Uexküll's theoretical biology to biosemiotics is the topic of another special issue of *Semiotica* to appear in 2001. There are three papers in *Biosemiotica II* which also deal with the (pre)history of biosemiotics: K. Kull's account of "Biosemiotics in the 20th century" (385-413), B.O. Brogaard's "Aristotelian approach to animal behavior" (199-213), and a paper by F. Stjernfelt on Kant's philosophy of biology and nature (537-566).

C. S. Peirce is undoubtedly the semiotician whose name is most frequently quoted (although not without misspellings: pp. 106, 213) by the authors of this volume. Peirce's specific contributions to the interface between semiotics and biology is the focus of two papers by Lucia Santaella. In "Peirce and biology" (5-21), she gives an outline of the locus of biology within Peirce's system of the sciences and of Peirce's research in the biochemical properties of protoplasm. Her second paper (497-519) deals with Peirce's theory of final causation in life, semiosis, biological, and cosmic evolution. Peirce's theory of evolution as 'habit taking,' his synechistic thesis of the continuity between matter and mind, his theory of the three categories, his definition of semiosis, and his typology of signs serve as a theoretical basis of many other papers, e.g., the ones by J. Hoffmeyer on "Order out of indeterminacy" (321-343) and by F. Merrell on "Living signs" (453-479).

---

[1] *Biosemiotica*, edited by Thomas A. Sebeok, Jesper Hoffmeyer and Claus Emmeche, Mouton de Gruyter, Berlin (=*Semiotica* 127.1/4 [Special Issue]), 1999, 658 pages.

[2] Winfried Nöth is the Director of the Research Center for Cultural Studies of the University of Kassel, Germany, and Visiting Professor at the Postgraduate Program in Semiotics and Communication Studies of the Catholic University of São Paulo, Brazil.

Not all contributors to this volume are equally familiar with the Peircean foundations of semiotics. One of them takes the hair-raising liberty of using Peirce to arrive at the following conclusion: "*Dyadic relationships*: The sign is the artifactual result of semiosic codification. By this I mean that it exists in the Peircean state of Secondness, which, as 'the mode of being of one thing which consists in how a second object is' (*CP* 1.224 [*recte*: *CP* 1.24, W.N.]) moves that energy into a distinct state-of-existence. This 'artifactual sign' is real as an encoded spatiotemporally unique entity" (599-600). – Nothing could be more against the semiotic spirit here evoked: the Peircean sign is certainly not constituted by a "dyadic," but only by a triadic relationship; it is not a phenomenon of Secondness, but only one of Thirdness; it is not a result of codification (see also p. 524 on evolution *without* coding), but of habit taking; it does not have its origin in energy (which in fact belongs to the category of secondness, of 'brute force,' of necessary, or efficient, causality), but it originates in final causation, which cannot be reduced to mere effects of energy. Last but not least, the Peircean sign is not necessarily artifactual at all, since it can also be manifested in the form of a mere thought or even a living human being; nor is the sign necessarily 'spatiotemporally unique,' which it is only when it functions as a sinsign (but not when it is a qualisign or a legisign).

Biosemiotica, as the Latin morphology of the term suggests, is an open field of research, without fixed boundaries, in which a plurality of approaches converge in the common interdisciplinary endeavor to explore the interfaces between biology and semiotics. Since the general editor refrains from *defining* the term and the research field of 'biosemiotics' (although such definitions can be found elsewhere, e.g., in Kull's paper, pp. 386, 388) and the guest editors resist the temptation of *structuring* this field by presenting the contributions in some kind of a systematic (instead of a merely alphabetical) order, the reader is free to discover the major and minor branches of biosemiotic tree of knowledge on his or her own. Topics which most probably constitute the major branches are: the role of semiosis in biological evolution in general (133-150, 481-496) and in human evolution in particular (227-238, 631-646), the relationship between meaning and information (237-284, 521-536, 599-612), the biological roots of intentionality (567-598), the material substratum of signs and semiosis (369-384), and finally cellular or genetic information, codification, and 'communication' (151-168, 273-284). Among the minor branches we find cultural theory and philosophy and of nature (345-368, 537-566) or such surprising newcomers to the semiotic field as 'literary biosemiotics' (239-272) and 'biohermeneutics' (215-226). The neighboring disciplines whose relevance to biosemiotics has been explored in this volume range from cybernetics (169-199) to linguistics (345-368) and psychoanalysis (613-630).

In this rhizomatic growth of biosemiotic research, it is not surprising that the plurality of approaches occasionally results in an astounding multiplication of semiotic or quasi-semiotic terminology. As long as these terms are duly defined, even neologisms such as *autokatakinetics* (a synonym of 'self-organization,' p.

584), *physicobiology* and *semiobiology* (the latter used as a synonym of 'biosemiotics,' p. 370), or even *intropy* and *enformation* (sic!, p. 489) must be tolerated, but incompatible with the ethics of terminology is the invention of the term *semetic* on the basis of its etymological explanation as "from Greek: semeion = sign, ethos = habit" (p. 342). If *ethos* is really to serve as the etymological root of this neologism, the form of this term must be *semethic*, but if the affix *-etic* is meant (in analogy to *phonetic* and the semiotic distinction between *etic* and *emic*), the root of this form is a mere adjectival derivative suffix, and it cannot be the Greek word *ethos*.

Readers of *Cybernetics & Human Knowing* will be especially interested in the papers dealing with aspects of the interfaces between cybernetics, dynamic systems theory, and biosemiotics. The most general panorama of these interrelationships is outlined in Søren Brier's paper with the baroque, but telling title "Biosemiotics and the foundation of cybersemiotics: Reconceptualizing the insights of ethology, second-order cybernetics, and Peirce's semiotics in biosemiotics to create a non-Cartesian information science" (169-198). Other papers deal with more specific topics of systems theoretical interest, such as self-reference (295-320, 524), autopoiesis (626), autocatalysis, and self-organization (523-25, 584), or semiotic (semantic, psychological) closure (524, 613-630).

The most innovative insights (or at least problems to think about) which biosemioticians will find in this volume seem to be those that pertain to the interface between biological and machine semiotics. Traditional biosemioticians used to determine the origins of semiosis at the threshold between matter and mind, physical and life sciences and emphasized that organisms "do *not* interact like mechanical bodies" (p. 386). Now, there is the challenge of machine semioticians who believe in the possibility of artificial semiosis and machine life, e.g., in cellular automata (295-320). Today there seems to be a generation conflict among the biosemioticians: whereas Thure von Uexküll, the senior scholar in the field of biosemiotics (b. 1908), writes against "mechanical models of explanation in the life sciences" (647-655) and "mechanistic ways of looking at things" (p. 647), younger contributors to *Biosemiotica II* are beginning to investigate the similarities between life and machines. With reference to Stuart Kauffman, Y. Kawade even predicts that in the foreseeable future, self-reproducing molecular systems will be created by humans so that "the traditional view that machines belong to the inanimate world no longer seems to be tenable" (p. 372-73).

However that may be, physical biology seems to provide an unexpected bridge between the tradition of theoretical semiotics and the avant-garde of machine semioticians. If "semiotic control" is a key concept in machine semiotics according to Howard Pattee (as quoted by Kull, p. 387) and such control is "local and conditional" in contrast to physical laws which are "global and inexorable" (ibid.), such an account of the threshold between physics and machine semiotics is in fact not incompatible with Peirce's theory of the difference between (physical) Secondness and (semiotic) Thirdness, efficient and final causation, or 'brute action' and the agency of the sign.

The Messenger

out of this windy gray icebox
through rain & clouded skies
through the Ides of March
the equinox awakens her brood
emerging, with no apologies
her message reels and turns
and jigs across the heavens
an astronomical hard fact, unfolding
racing toward Andromeda
steadily marching up the hemisphere
reflected in every bit of light
from every candelabra
and every candle stick
in all the forms of life
cradling the negentropic fires
in this known region
of the cosmos